Robert C. Brears
Sustainable Water-Food Nexus

Also of interest

Water Resources Management.
Innovative and Green Solutions
Robert C. Brears, 2024
ISBN 978-3-11-102807-1, e-ISBN 978-3-11-102810-1

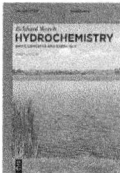

Hydrochemistry, 2nd Edition.
Basic Concepts and Exercises
Eckhard Worch, 2023
ISBN 978-3-11-075876-4, e-ISBN 978-3-11-075878-8

Drinking Water Treatment.
New Membrane Technology
Bingzhi Dong, Tian Li, Huaqiang Chu, Huan He,
Shumin Zhu and Junxia Liu (Eds.), 2021
ISBN 978-3-11-059559-8, e-ISBN 978-3-11-059684-7

Water Resource Technology.
Management for Engineering Applications
Vikas Dubey, Sri R.K. Mishra, Marta Michalska-Domańska
and Vaibhav Deshpande (Eds.), 2021
ISBN 978-3-11-072134-8, e-ISBN 978-3-11-072135-5

Robert C. Brears

Sustainable Water-Food Nexus

Circular Economy, Water Management, Sustainable
Agriculture

DE GRUYTER

Robert C. Brears
Sustainable Water-Food Nexus

ISBN 978-3-11-134128-6
e-ISBN (PDF) 978-3-11-134138-5
e-ISBN (EPUB) 978-3-11-134147-7

Library of Congress Control Number: 2024949413

Bibliographic information published by the Deutsche Nationalbibliothek
The Deutsche Nationalbibliothek lists this publication in the Deutsche Nationalbibliografie;
detailed bibliographic data are available on the internet at http://dnb.dnb.de.

© 2025 Walter de Gruyter GmbH, Berlin/Boston, Genthiner Straße 13, 10785 Berlin
Cover image: Ratsanai/iStock
Typesetting: Integra Software Services Pvt. Ltd.

www.degruyter.com
Questions about General Product Safety Regulation:
productsafety@degruyterbrill.com

Acknowledgments

First and foremost, I wish to thank the team at De Gruyter who are visionaries and enable books like mine to come to fruition. Second, I would like to express my gratitude to my mum, who has a keen interest in environmental matters and has supported me in this journey of writing the book. Lastly, I want to extend a special acknowledgment to Kate, my love, who enriches my life in countless ways and has been an inexhaustible source of inspiration and motivation during the creation of this work.

https://doi.org/10.1515/9783111341385-202

Contents

Tables and cases

https://doi.org/10.1515/9783111341385-204

Chapter 1
Introduction

Abstract: This chapter introduces the critical challenges of rising global demands for water and food, emphasizing the need for sustainable resource management within the framework of a circular economy. It provides an overview of the water-food nexus, highlighting the interconnectedness of water and food systems and the environmental pressures resulting from traditional linear economic models. It sets the stage for exploring how circular economy principles, such as reducing waste and maximizing resource efficiency, can be applied to achieve sustainable water and food management. The chapter also outlines the 5R framework (reduce, reuse, recycle, recover, and restore) as a foundational guide for managing resources and introduces key themes and solutions that will be discussed throughout the book.

1.1 Introduction

As global demands for water and food continue to rise, the need for sustainable resource management has become increasingly urgent. This book offers an in-depth exploration of the interconnected relationship between food and water systems within the framework of a circular economy – a model focused on reducing waste and maximizing resource efficiency. It begins by defining the key concepts of the water-food nexus and the circular economy, setting the stage for a detailed discussion of their interrelated roles in sustainable development.

The book opens with an overview of the water-food nexus, emphasizing the essential role of water in food production and the broader implications for society. It critically examines how traditional linear economic models have contributed to environmental challenges and discusses the transition to a circular economy as a necessary step toward more sustainable resource use. The circular economy's 5R framework – reduce, reuse, recycle, recover, and restore – is a foundational guide throughout the text, illustrating how these principles can be applied to achieve sustainable water and food management.

Chapter 2 introduces the concept of the circular economy and contrasts it with the linear economy, highlighting the benefits of circular practices. Within this context, the chapter explores agriculture's pressures on water resources and discusses strategies to alleviate these pressures. Practical examples, such as water conservation initiatives and sustainable agricultural practices, demonstrate how circular principles can be applied in real-world scenarios.

Chapter 3 delves into enhancing water efficiency and conservation within agriculture, exploring how the circular economy approach can optimize water use and reduce waste. The chapter covers various water-saving technologies and practices, in-

https://doi.org/10.1515/9783111341385-001

cluding drip irrigation, soil moisture sensors, and drought-resistant crops, and discusses their potential to improve agricultural water management.

Chapter 4 examines sustainable agriculture and nature-based solutions, focusing on their potential to maintain water quality and reduce the environmental impact of conventional farming practices. The chapter discusses innovative techniques, such as precision farming and integrated pest management, which aim to optimize water use and enhance soil health, contributing to more sustainable agricultural systems.

Chapter 5 focuses on water reuse and recycling within circular food systems. It explores the technologies and practices that enable water reuse and recycling in agriculture and food processing, offering examples of how these methods can conserve resources and reduce waste. The chapter also highlights successful case studies, where treated wastewater and reclaimed water have been effectively used for irrigation and industrial purposes.

Chapter 6 addresses the interconnections between food production, water use, and energy consumption, emphasizing the importance of harmonizing these elements to promote sustainability in agriculture. The chapter discusses optimizing resource use within the food-water-energy nexus, including integrating renewable energy sources such as solar and wind power to reduce the carbon footprint of food production.

Chapter 7 explores the growing trend of urban agriculture as a component of circular economy practices within metropolitan areas. This chapter examines how cities can implement innovative solutions for water management and local food production, highlighting successful projects that integrate vertical farming, hydroponics, and aquaponics. These approaches are presented as viable strategies for enhancing food security, reducing food miles, and supporting sustainable urban development.

Chapter 8 investigates the role of financial investments and strategic partnerships in facilitating the transition to a circular economy within the water-food nexus. The chapter analyzes various financing mechanisms, including public and private funding, and discusses how these resources can support the scaling up of circular practices and technologies. Case studies and examples of innovative financing approaches illustrate the potential for overcoming financial barriers and advancing sustainable resource management.

Chapter 9 identifies best practices from case studies throughout the book and presents key conclusions on implementing circular economy principles within the water-food nexus. The chapter highlights successful strategies, technologies, and partnerships that have driven sustainable water and food management. Through an analysis of case studies, it draws out actionable insights that can be applied to future projects. The chapter concludes by summarizing the overall lessons learned and providing recommendations for stakeholders to continue advancing circular practices, fostering innovation, and ensuring the resilience of the water-food nexus.

This book serves as a valuable resource for researchers, policymakers, and practitioners interested in advancing sustainable resource management. Through its thorough exploration of the water-food nexus within the context of the circular economy, the book provides practical insights and guidance for fostering a more sustainable and resilient future.

Chapter 2
Defining the circular economy and the water-food nexus

Abstract: This chapter examines the intersection of the circular economy with the water-food nexus, emphasizing the need to transition from a linear to a circular economic model to address environmental sustainability challenges. It outlines the water-food nexus's complexities, focusing on water's critical role in society and the importance of food security. The discussion introduces the circular economy as a strategy to reduce waste, maintain product utility, and regenerate natural systems. It offers examples to demonstrate its application in mitigating water-food nexus pressures. The chapter advocates for the circular economy's 5R approach (reduce, reuse, recycle, recover, and restore) for sustainable water and food management.

2.1 Introduction

In our traditional economic system, three types of capital – manufactured, human, and natural – play vital roles in human welfare by aiding in the production of goods and services. Natural capital, which includes the Earth's unprocessed natural resources, is pivotal for providing material and energy for production and absorbing waste from economic activities. This system is typically linear: economic actors, encompassing individuals or organizations involved in production, distribution, consumption, and resource management, extract natural resources to create products. These products are sold and eventually discarded by consumers when they cease to be useful. A significant issue in this linear economy is the increasing stress on the interconnection between water resources and food production, the water-food nexus.

This chapter will first provide an overview of the components of the water-food nexus and related trends before introducing the circular economy concept. The chapter will then discuss how water managers, following the circular water economy principles of reduce, reuse, recycle, recover, and restore, implement policy innovations to reduce water-food nexus pressures.

2.2 Components of the water-food nexus

The individual components of the water-food nexus are as follows.

https://doi.org/10.1515/9783111341385-002

2.2.1 Water

Like any other resource, water is integral to all aspects of society and the environment and is crucial for human well-being. It plays a vital role in food security, health, and poverty reduction and drives economic growth in agriculture, industry, and energy generation. Transitioning to a circular economy thus necessitates conserving water resources and exploring new, sustainable economic and social development opportunities through effective water management. A fundamental aspect of building the circular economy is ensuring water security. This involves guaranteeing that all human and ecological users have sustainable access to sufficient, high-quality water. This access is essential for maintaining livelihoods, human well-being, socioeconomic development, protecting against water pollution and related disasters, and conserving ecosystems within peace and political stability. In the context of the circular economy, water security can be effectively established by:

-   implementing policy measures that foster benefits across economic, environmental, and social dimensions;
-   introducing fiscal tools that assign value to environmental assets;
-   enhancing institutional frameworks to facilitate water management across different sectors and beyond political or administrative borders;
-   developing financial strategies that distribute risks between governments and investors makes innovative water technologies accessible and economical;
-   cultivating skills essential for sustainable water management practices;
-   setting up robust information and monitoring systems to establish targets, outline paths, and track advancements in water efficiency; and
-   formulating innovative strategies to enhance water productivity, safeguard groundwater and surface water resources, and maintain optimal water quality standards.

2.2.2 Food

The food sector has effectively met the growing global demand for an extended period. Agricultural productivity has surpassed many other sectors and outpaced population growth. Besides providing food, agriculture can significantly reduce poverty in developing countries, impacting income growth for the poorest at least 2.5 times greater than nonagricultural sectors. In the shift toward the circular economy, the food and agricultural sector aims to minimize environmental impacts while boosting productivity and farmer incomes, ensuring food security for all. The Food and Agriculture Organization defines food security as the condition where "all people at all times have physical, social, and economic access to sufficient, safe, and nutritious food to meet their dietary needs and food preferences for an active, healthy life." A critical aspect of achieving food security is acknowledging that while intensifying crop production can enhance the food security of millions globally, it can also lead to land degradation, water pollution, and depletion of water resources, posing risks to food security. In the circular economy, food security can be attained through farming practices and technologies that both maintain and increase farm productivity and profitability and ensure the sustainable provision of food and ecosystem services. These practices should aim to reduce negative externalities until positive outcomes are achieved and rebuild ecological resources such as soil, water, air, and biodiversity by reducing pollution and using resources more efficiently.

2.3 Interaction of water and food: the nexus

Currently, 47% of the global population, or 3.6 billion people, experience water scarcity for at least 1 month each year, a figure projected to rise to 57% by 2050. Over the past century, global water demand has surged by 600%, with forecasts indicating a 20–30% increase in demand across industrial, domestic, and agricultural sectors by 2050, pushing total water use from the current 4,600 km^3/year to between 5,500 and 6,000 km^3. Despite faster growth in industrial and domestic sectors, agriculture, which presently accounts for 70% of water use primarily for irrigation, will see the largest absolute increase in demand, expected to grow by 60% by 2025. This increase is in response to the need to boost food production by 60% by 2050 to meet global demand, necessitating expanded arable land and intensified production methods, further elevating water usage. Groundwater, a critical resource for agriculture, is already being extracted at a rate of 800 km^3/year as of the 2010s, with projections showing an increase to 1,100 km^3 by 2050, a 39% rise. Over 30% of the world's largest groundwater systems are currently overstressed, with many regions lacking precise data on the remaining water reserves, leading to unsustainable consumption rates without knowledge of when these crucial reserves might be depleted [1, 2].

Climate change is set to significantly deepen agricultural water scarcity, impacting more than 80% of global croplands. The main driver of this increasing scarcity is the anticipated decrease in water availability, which will be the dominant factor in exacerbating water scarcity issues worldwide. Furthermore, the role of green water – soil moisture used by crops – is expected to contribute substantially to the changes in water scarcity conditions, particularly in 16% of global croplands [3].

The rise in agricultural production will adversely affect water quality through nonpoint source pollution (Table 2.1). Key issues include sediment runoff leading to siltation, nutrient runoff from nitrogen and phosphorus in fertilizers, animal manure, and municipal wastewater, microbial runoff from livestock or using excreta as fertilizer, and chemical runoff from pesticides and other agrichemicals contaminating water sources.

Table 2.1: Agricultural impacts on water quality.

Agricultural activity	Impacts on surface water	Impacts on groundwater
Tillage/ ploughing	Sediments and turbidity carry phosphorus and pesticides; siltation of riverbeds and loss of habitat	Tillage increases sediment and pollutant load, potentially contaminating groundwater
Fertilizing	Runoff of nutrients, including phosphorus, leads to eutrophication	Leaching of nitrate
Manure spreading	High levels of contamination by microorganisms, phosphorus, and nitrate, causing eutrophication	Contamination by nitrogen
Pesticides	Runoff of pesticides contaminates surface water, affecting ecological systems	Potential leaching of pesticides, and human health risks from contaminated wells
Feedlot	Contamination with microorganisms, veterinary drug residues, and metals in urine and feces	Potential leaching of nitrogen, metals, etc.
Irrigation	Runoff of salts causes salinization, and ecological damage from fertilizers and pesticides	Contamination with salts and nutrients
Clear-cutting	Land erosion leads to turbidity and siltation. Hydrological changes with loss of streams	Disruption of hydrological regime, increased surface runoff, and decreased recharge

Irrigation salinity poses a significant risk by raising the water table and mobilizing soil salts, which, when combined with waterlogging, hinder plant growth by limiting oxygen access and disrupting water and nutrient uptake. This problem is exacerbated by factors like heavy rainfall, leading to salt accumulation that prevents plants from effectively absorbing water and nutrients.

Similarly, fertilizer runoff contributes to the eutrophication of water bodies, a process accelerated by excessive nitrogen and phosphorus, rendering these waters unsuitable for various uses. Eutrophication encourages the overgrowth of algae and aquatic weeds, affecting agriculture, recreation, and water quality. In marine environments, algal blooms can cause shellfish poisoning and create oxygen-depleted "dead zones," unable to support aquatic life. Freshwater cyanobacteria blooms, influenced by nutrient runoff, result in fish kills, tainted drinking water, and health risks to livestock and humans, including conditions like "blue baby disease" caused by nitrate contamination [4].

An overall summary of the risks and impacts of the water-food/food-water nexus is summarized in Table 2.2.

Table 2.2: Summary of risks and impacts in the water-food/food-water nexus.

Risk factor	Impact on water security	Impact on food security	Notes
Water scarcity	Increased competition for water resources; depletion of freshwater reserves	Fluctuations in food supply; increased irrigation demand	Climate change exacerbates scarcity, impacting 80% of global croplands
Groundwater overuse	Depletion and stress on 30% of the world's largest groundwater systems	Compromises sustainable agriculture due to uncertain water reserves	Groundwater extraction expected to rise by 39% by 2050
Agricultural demand	Elevated water withdrawals for irrigation; strain on water reserves	Necessity for expanded arable land and intensified production methods	Agriculture accounts for 70% of water use; demand expected to grow by 60% by 2025
Climate change	Decreased water availability; over 80% of croplands affected by scarcity	Requirement for 60% increase in food production by 2050	Significant contributions from green water scarcity in 16% of global croplands
Nonpoint source pollution	Eutrophication; salinization; contamination by pesticides and nitrates	Soil degradation; reduced crop yields; health risks from contaminated food	Impacts from tillage, fertilizing, manure spreading, and pesticide runoff
Irrigation practices	Mobilization of soil salts; waterlogging; decreased water quality	Hindered plant growth due to oxygen access and nutrient uptake disruption	Exacerbated by heavy rainfall and poor water management strategies

2.4 Circular economy

The circular economy represents a transformative approach to production and consumption, contrasting with the traditional linear model that depends on a cycle of extracting, using, and discarding resources. It is a strategic response to the environmental pressures of our time, aiming to redefine growth by decoupling economic activity from the consumption of finite resources. Its principles are straightforward yet profound: eliminate waste and pollution through innovative design, retain the utility and value of products and materials perpetually, and regenerate natural systems.

In contrast to the current system that recycles a mere fraction of materials – currently, only 7.2% of used materials are cycled back into our economies after use – the circular model envisions a closed-loop economy where everything is reused and sustained. It is a comprehensive framework that tackles the underlying causes of environmental crises, including climate change and biodiversity loss, by fostering a culture of conservation, longevity, and resourcefulness. It addresses pollution and presents a viable solution to other complex environmental challenges.

The ethos of the circular economy includes extending the lifespan of products through sharing, repairing, refurbishing, and recycling. It encourages a shift from ownership to access, suggesting consumers lease or share what they need, reducing the demand for new products. This model promotes a systematic redesign of the life cycle of products, ensuring that at their end of life, materials are not wasted but are reintroduced into the economy, maintaining their value and reducing the need for new raw materials.

The circular economy fosters an industrial system aligned with ecological principles by emphasizing renewable energy and materials. It moves away from planned obsolescence, where products are designed for a limited lifespan, a practice now scrutinized for its environmental impact. The circular approach offers environmental, economic, and social benefits, creating opportunities for innovative business models and sustainable practices that can lead to better growth opportunities and a more resilient economy [5–9].

2.4.1 Principles of the circular economy

The circular economy operates on three main principles, each aimed at creating a sustainable and waste-free cycle of production and consumption:
- Eliminate waste and pollution: This principle focuses on redesigning the life cycle of products to prevent waste and pollution from the outset. It involves creating products that can be fully reused, repaired, or recycled, thereby extending their lifespan and usefulness. The idea is to transition from the linear model of "take-make-waste" to one where products and materials are perpetually cycled through the economy with minimal environmental impact.
- Circulate products and materials at their highest value: The second principle emphasizes keeping materials in use at their highest utility. This involves designing for durability and longevity and implementing systems for repair, refurbishment, and remanufacturing. In this way, materials are kept within the economic loop, retaining their value and reducing the need to extract and process raw materials.
- Regenerate natural systems: Moving beyond merely reducing harm, the circular economy actively seeks to improve and regenerate the natural environment. This involves embracing regenerative agriculture, which rebuilds soil health and biodiversity. It also means designing ways to return organic materials to the Earth in a manner that supports the ecosystem, turning waste into a resource.

These principles aim to transform our economic system into one that mirrors natural processes, where nothing is wasted, and everything has value. By applying these principles, the circular economy seeks to create a sustainable, resilient, and prosperous future [8].

2.4.2 Benefits of the circular economy

The circular economy represents a significant shift toward sustainability, with potential environmental, economic, and social benefits:
–   *Environmental protection*: A circular economy aims to decrease natural resource use and habitat destruction, thereby preserving biodiversity and lowering greenhouse gas emissions.
–   *Resource independence*: It reduces reliance on raw materials; some of which are scarce or require importation, thus mitigating the risk of supply shortages and trade deficits.
–   *Economic and consumer benefits*: This model can fuel economic growth and innovation, potentially creating numerous jobs and saving consumers money through more durable products.
–   *Efficient resource use*: The circular economy maximizes the utility of forests, minerals, and other natural assets by reusing resources.
–   *Emission reduction*: Circular strategies can significantly cut global emissions, contributing to climate change mitigation.
–   *Health and biodiversity*: By minimizing pollution, the circular economy safeguards human health and protects the diversity of life on the Earth.
–   *Economic growth*: It is projected to offer a multitrillion-dollar economic opportunity by driving innovation, reducing waste, and creating jobs.
–   *Job creation*: The transition could result in millions of new jobs, although it necessitates a just transition for workers from linear industries to circular ones [7, 10].

2.4.3 Enacting the circular economy

A more circular world would require system change. Consumers, businesses, and politicians must change how goods are designed, produced, sold, manufactured, and reused. Some shifts could include:
–   Design: In a circular economy, product design is critical and emphasizes sustainability. The goal is to design products that use fewer raw materials and are easily disassembled for repair, reuse, or recycling. This approach promotes longevity and resource efficiency, aligning with principles that reduce the environmental footprint of manufactured goods.
–   Business models: Product-as-a-service and sharing economy models are contemporary business strategies that integrate sustainability by allowing customers to access services or share ownership rather than buying products outright. These models prioritize the efficient use of resources and support a shift toward more sustainable consumption patterns.
–   New metrics and shared data: Enhanced data on material availability could significantly cut natural resource use in traditionally challenging industries to make

sustainable. Establishing standardized metrics is vital to this effort, as it would provide a consistent basis for measuring reuse and its impact on reducing the strain on natural resources. This step is fundamental in addressing broader environmental issues like resource depletion and fostering a shift toward more resource-efficient models.

- Policy: Incentivizing businesses and communities to adopt circular practices is crucial for meaningful, sustained environmental progress. Effective incentives can drive the adoption of reduction, reuse, and recycling habits, ultimately leading to a significant decrease in waste and more sustainable resource use. Legislation plays a pivotal role, with policies and regulations steering the collective toward a circular economy by setting ambitious, actionable targets for resource conservation and waste reduction [10].

2.5 Circular water management within the water-food nexus

Following the principles of the circular economy is essential to reduce water-food nexus pressures. The 5R framework – reduce, reuse, recycle, recover, and restore – provides a structured approach to optimizing water use, minimizing waste, and managing water resources sustainably within agricultural systems:

- Reduce: Employ water-efficient techniques, such as precision irrigation and optimized crop management, to lower overall water demand and reduce pressure on water resources.
- Reuse: Encourage the use of reclaimed water, including treated wastewater, in agricultural and food processing activities to decrease dependence on freshwater supplies.
- Recycle: Utilize water recycling technologies to treat and repurpose agricultural runoff and gray water, extending the usability of water within the system.
- Recover: Extract nutrients from wastewater and agricultural effluents for use as fertilizers, supporting soil health and reducing the need for synthetic inputs.
- Restore: Apply nature-based solutions to restore natural water cycles, recharge aquifers, and support ecosystem services essential for maintaining the balance within the water-food nexus [11, 12].

2.5.1 Reduce

Water reduction is crucial in agriculture. Applying circular economy principles can significantly impact water conservation. Water conservation involves behavioral changes, like altering irrigation practices during droughts to minimize water use. Sustainable water management in agriculture is vital for increasing production, sharing water with other users, and maintaining the benefits of water systems. Water efficiency involves optimizing the benefits of each water unit used. It often pairs existing needs with technological improvements to achieve the same results with less water. For example, precision irrigation techniques, such as drip irrigation, allow water to drip slowly directly to the crop roots, enabling precise water application without waste due to excessive evaporation or imprecise coverage. Advanced technologies like soil moisture sensors and satellite evapotranspiration measurements can further improve water efficiency and productivity in agriculture. Digital tools provide farmers with real-time field information, ensuring water is used only when and where needed. Applying circular economy principles to agricultural water usage encourages both conservation and efficiency. This helps preserve water resources and contributes to agricultural sector sustainability. It benefits both the environment and the economy [12–15].

Case 2.1 Reduce: Australia's Drought Resilience Innovation Grants

The Drought Resilience Innovation Grants, part of the Australian Department of Agriculture's Future Drought Fund, aim to enhance drought resilience through innovative solutions in agriculture. Announced in July 2021, these grants are categorized into Ideas Grants ($50,000 for 12 months), Proof-of-Concept Grants (up to $120,000 for 12 months), and Innovation Grants (between $300,000 and $1.1 million annually for up to 3 years). A key focus of these grants is water conservation. For example, one project funded by Ideas Grants is developing a slow-release fertilizer pellet to improve soil moisture retention, demonstrating a commitment to innovative water management in agriculture. Another notable project under the Innovation Grants explores capturing and reusing near-shore groundwater for irrigation, showcasing advanced solutions in water conservation. These grants are pivotal in fostering new technologies and strategies for water management, which are crucial for the sustainability of Australian agriculture in the face of frequent droughts [16].

Case 2.2 Reduce: Arizona's Water Irrigation Efficiency Program

The Water Irrigation Efficiency Program in Arizona, launched in 2023, aims to reduce water usage in the state's agricultural sector, which accounts for 78% of Arizona's total water demand. Funded by the state and administered by the University of Arizona, the program encourages farmers to switch from traditional flood irrigation methods to more efficient systems like drip and sprinkler irrigation. The goal is to cut water usage by at least 20% across participating farms. Farmers enrolled in the program receive $1,500 per acre, with a maximum of $1 million per farm, to assist in purchasing approved irrigation systems. By the end of 2023, the program had saved 36,418 acre-feet of water, significantly exceeding expectations. The program targets regions where water resources are especially scarce, such as central and southern Arizona, aiming to preserve groundwater and Colorado River water supplies. In addition to financial incentives for farmers, the program also allocated $2 million for research to improve irrigation techniques and develop more effective methods for measuring soil moisture. This research component seeks to enhance long-term water conservation strategies and support sustainable agricultural practices [17].

2.5.2 Reuse

Water reuse is a practice in which water is used more than once within the same process or for different purposes without treatment. This practice is gaining importance due to the increasing competition for water resources and the emergence of alternative water sources. During droughts, when conventional water supplies are scarce or unavailable, reusing water becomes a critical resource in agriculture. Water reuse involves treating domestic or municipal wastewater, gray water, return flows and tail-water, produced water, brackish water, and others for specific uses, like crop irrigation before it reenters the natural water cycle. Technologies such as filtration systems and treatment processes are often employed to purify reclaimed water to meet regulatory requirements. Reusing water for agricultural irrigation can help reduce costs associated with water importation, reduce freshwater demands, and create a reliable, sustainable, and local water supply. Overall, water reuse in agriculture is a beneficial practice that promotes water conservation, especially during droughts, and contributes to the efficient use of water resources. It not only helps in preserving water resources but also aids in maintaining agricultural productivity [12–15].

Case 2.3 Reuse: guidelines for Sustainable Use of Greywater in Small-Scale Agriculture and Gardens in South Africa
In South Africa, gray water reuse in agriculture is emerging as a viable solution to address water scarcity and reduce stress on freshwater sources. The Water Research Commission has developed guidelines to facilitate the sustainable use of gray water for garden and small-scale agricultural irrigation. Gray water, which is untreated household wastewater excluding toilet waste, primarily comes from baths, showers, kitchen sinks, and laundry. This source accounts for over half of a household's indoor water use and offers a significant opportunity for reuse. Currently, gray water use is practiced informally, both in urban gardens of middle- to upper-income areas and in food gardens of lower-income, periurban, and rural communities. This practice is particularly beneficial in supplementing irrigation water and providing nutrients for crop plants, thereby contributing to food security in poorer communities. Despite its growing use, formal guidelines for gray water utilization in South Africa are not yet established [18].

Case 2.4 Reuse: Abu Dhabi's New Water Source

The Environment Agency – Abu Dhabi (EAD) prioritizes the preservation and sustainability of water resources, addressing the growing demand for freshwater in the Emirate of Abu Dhabi. In this context, the EAD is exploring nonconventional water sources, mainly focusing on water produced from oil and gas fields. This initiative involves extensive research into the potential for treating and reusing this specific type of water. Collaborating with various partners, the EAD has undertaken a comprehensive study to assess the quantity and quality of water extracted during oil and gas extraction, evaluating suitable methods for its management, treatment, and potential reuse. The study reveals that approximately five barrels of water are extracted from every barrel of oil produced in Abu Dhabi. This water, characterized by high salinity and a complex composition of mineral salts, organic compounds, hydrocarbons, organic acids, oils, greases, and other chemicals, requires rigorous treatment before reuse. The treatment processes include membrane technology, thermal methods, hydrocyclones, biological aerated filters, adsorption, enhanced flotation, and ion exchange, each aiming to eliminate impurities and desalinate the water. Despite the high economic costs, these treatment methods are environmentally friendly. The treated water can be reused in various ways, such as in oil and gas industries, reinjection into oil wells, discharging into marine environments, irrigation, or even in creating wetlands, provided that it meets environmental standards [19].

2.5.3 Recycle

In agriculture, water recycling means the harvested water is treated and reused. The treatment level can vary based on the intended use. Primary treatment removes suspended solids and organic matter. Secondary treatment uses microorganisms to eliminate organic matter and suspended material. Tertiary treatment involves the advanced removal of suspended and dissolved materials through chemical disinfection and filtration. Recycling water in agriculture has several advantages. It provides a consistent water supply enriched with nutrients and organic matter, enhancing crop production. It also reduces dependency on traditional water sources, mitigating the impacts of seasonal variations and climate change on water availability. Furthermore, it promotes environmental conservation by minimizing wastewater discharge into natural bodies of water, thus preserving aquatic ecosystems. It also reduces the energy footprint associated with traditional water supply methods. In summary, water recycling in agriculture is a beneficial practice that promotes water conservation and efficiently uses water resources. It not only helps preserve water resources but also aids in maintaining agricultural productivity [12–15].

Case 2.5 Recycle: Recycled Water for Agricultural Irrigation in Kent County, Maryland

In Kent County, Maryland, a partnership between local farmers and the Worton-Butlertown Wastewater Treatment Plant (WWTP) has enabled the use of recycled water for agricultural irrigation. This initiative addresses water scarcity by utilizing highly treated municipal wastewater to irrigate crop fields. The treatment plant, using advanced processes such as membrane filtration and ultraviolet light, provides effluent water that meets strict quality standards set by the Maryland Department of the Environment. These standards ensure safe levels of turbidity, pH, and nitrogen, and require continuous monitoring and reporting. The program's implementation involved significant infrastructure investments, including a $3 million irrigation system funded by county, state, and grant resources. Farmers participating in the initiative gain a reliable and regulated water source for their crops, especially in areas where rainfall is unpredictable. The project allows farmers to control when and how much water is applied, enhancing crop growth during dry periods. The use of recycled water has already demonstrated benefits in increasing crop yields. Cornfields, for example, have seen an increase of 30–100 bushels per acre, depending on rainfall, and soybean yields have improved by 11–12 bushels per acre during dry years [20].

Case 2.6 Recycle: recycling treated wastewater at the Al Wathba-2 WWTP in Abu Dhabi

The Al Wathba-2 WWTP in Abu Dhabi is integral to the Emirate's efforts to manage water scarcity by recycling treated wastewater. In a region with minimal renewable freshwater resources, the plant contributes significantly to Abu Dhabi's integrated water resource management plans, focusing on sustainability and reducing environmental impact. The plant processes wastewater to a tertiary level, producing approximately 105,000 m^3 of recycled water daily. This water is primarily allocated for irrigating landscaped areas, green spaces, and agricultural lands, reducing dependence on desalinated water. Using recycled water instead of desalinated water helps decrease the energy consumption associated with desalination and lowers carbon dioxide emissions. The production cost of recycled water is approximately $0.051/$m^3$, compared to $2.77/$m^3$ for desalinated water, making recycled water more economical. The Al Wathba-2 WWTP has installed biotrickling filters in its wastewater pumping stations to improve environmental outcomes. These filters, which replace traditional chemical scrubbers, are designed to remove odorous gas compounds from the recycled water, ensuring that the water meets quality standards for reuse. In addition, the plant's recycled water is used for ecological applications, such as irrigating the Al Wathba Wetlands. In 2022, it was projected that an additional 390,000 m^3/day of recycled water from the Al Wathba-1 and -2 WWTPs would be used to irrigate 4,200 farms, supporting both food production and environmental sustainability in the region. Reusing treated wastewater also replenishes deteriorated groundwater aquifers, improves water quality, and increases reserves for future use [21].

2.5.4 Recover

In agriculture, recovery is embodied using biosolids, nutrient-rich organic substances derived from treated sewage sludge. The U.S. Environmental Protection Agency (US EPA) recognizes biosolids that have undergone significant processing to reduce pathogens and odors, making them safe for agricultural application. The use of biosolids in farming enhances soil productivity and sustains essential nutrients. They offer a controlled release of nitrogen, supporting plant growth throughout the season. Biosolids are an alternative to chemical fertilizers, providing a slow-release nutrient source for crops. This not only contributes to the health and productivity of the crops but also promotes sustainable farming practices. In conclusion, the recovery aspect in agriculture, represented by biosolids, contributes to the efficient use of resources, enhances soil productivity, and supports sustainable agricultural practices [12–15].

Case 2.7 Recover: recovering nutrients through biosolids production at DC Water's Blue Plains Advanced WWTP

DC Water's Blue Plains Advanced WWTP in Washington, D.C., is actively involved in the production of Class A Exceptional Quality biosolids, a nutrient-rich material derived from treated sewage sludge. The plant employs a sophisticated process that includes thermal hydrolysis and anaerobic digestion (AD) to treat the sludge. During this process, the solid waste is subjected to heat, pressure, and beneficial bacteria, which effectively eliminate pathogens and significantly reduce odors. The final product is a biosolid that meets US EPA standards, making it safe and effective for use in agriculture. The biosolids produced at Blue Plains are utilized across a variety of settings, ranging from large-scale farms in Maryland and Virginia to community gardens and landscaping projects within Washington, D.C. These biosolids provide a sustainable alternative to conventional chemical fertilizers by recycling essential nutrients like nitrogen and phosphorus back into the soil. This helps improve soil fertility, support plant growth, and enhance overall soil health. The slow-release nature of the nutrients in biosolids ensures a consistent supply to crops throughout the growing season, which is essential for maintaining productivity while minimizing the environmental impact associated with synthetic fertilizers. In addition, the use of these biosolids contributes to carbon sequestration efforts, as the organic matter in biosolids helps capture carbon in the soil [22].

Case 2.8 Recover: United Utilities implementing biosolids recycling in agriculture

United Utilities, a UK-based water company, annually recycles 360,000 tons of biosolids for agricultural use across 1,500 farms covering 18,000 hectares. Accredited by the Biosolids Assurance Scheme (BAS), United Utilities promotes recycling in agriculture. Biosolids, sewage sludge treated for agricultural land use, are endorsed by the UK Government and European Union for their environmental benefits. United Utilities has achieved 100% compliance for 5 years, ensuring biosolids meet required standards. Various treatment technologies, including advanced AD and AD, are used and closely monitored. United Utilities' Agriculture Advisors assist farmers in cost-effective fertilizer use and soil improvement. The company prioritizes safe and sustainable biosolids, and recycling is certified by the BAS, and is independently audited by NSF Certification UK Limited. United Utilities takes pride in its 100% BAS-certified biosolid products [23].

2.5.5 Restore

In agriculture, the concept of restoration is embodied in the optimization of environmental flows and the implementation of nature-based solutions:

- Environmental flows: Environmental flows refer to the specific water levels needed to maintain ecosystems and the human benefits derived from them.[13] In the circular economy, these are optimized by minimizing water usage, maintaining natural water quality, and reducing human impact on water bodies.
- Nature-based solutions: Nature-based solutions, such as the restoration of forests, grasslands, natural wetlands, soil conservation, and sustainable management of aquifers, further enhance the circular economy. These solutions help restore the natural balance of ecosystems, thereby supporting biodiversity and enhancing the resilience of agricultural systems.
- Water reuse and recycling: Strategies like water reuse and recycling help meet human water needs while ensuring more water remains for natural environments. This supports crucial ecosystems and contributes to a more sustainable and resilient water management system [12–15].

Case 2.9 Restore: restoring agricultural sustainability through conservation grants and incentives in Maryland

The Maryland Department of Agriculture offers a range of grants and loans to assist farmers in implementing best management practices (BMPs) that protect natural resources and comply with federal, state, and local environmental regulations. Through the Maryland Agricultural Water Quality Cost-Share Program, farmers can receive grants covering up to 100% of the costs associated with installing BMPs to control soil erosion, manage nutrients, and protect water quality in streams, rivers, and the Chesapeake Bay. Currently, around 40 BMPs are eligible for funding. The Cover Crop Program provides grants to farmers for planting small grains in the fall, helping to conserve nutrients, reduce soil erosion, and safeguard water quality. The Conservation Reserve Enhancement Program, a federal-state partnership, offers landowners attractive rental rates to remove environmentally sensitive cropland from production for 10–15 years, during which they plant buffers and other conservation practices that enhance water quality and provide wildlife habitat. Additionally, the department offers incentives for tree planting, which can be explored further through their tree planting incentives. The Manure Transport Program supports animal producers in moving excess manure off their farms, while cost-share grants for manure injection are available in both spring and fall to help farmers comply with Maryland's nutrient management regulations. Farmers purchasing specific conservation equipment may qualify for a Maryland Income Tax Subtraction Modification. The Low Interest Loans for Agricultural Conservation (LILAC) program supplements federal and state cost-share payments, and grants are also available for developing innovative animal waste technologies [24].

Case 2.10 Restore: sustainable farming practices in the Mulkear Catchment Limerick and Tipperary counties of Ireland
Agricultural production in Ireland has significantly increased in the past 5 years, leading to a growing need for sustainable farming practices to alleviate pressure on Ireland's natural water bodies. The Mulkear Operational Group project, led by Project Manager Carol Quish, supported farmers in mitigating agricultural impacts and enhancing water quality within the "at-risk" Mulkear catchment in Ireland's Limerick and Tipperary counties. The project designed a program to support farming activities while improving water quality at the catchment scale – the local, farmer-led collaborative partnership program aimed to codesign an innovative suite of mitigation measures. Farmers in the program attended regular discussion group meetings to identify problems and cocreate practical and sustainable water management solutions. The project team worked with each farmer to develop a plan with specific measures to improve water quality, such as nutrient management, grazing infrastructure, and farmland enhancement. One of the measures applied was low-emission slurry spreading, which provided a more efficient use of nitrogen while protecting the environment. In 2022, the project team assessed the success of the different mitigation measures in improving water quality. It provided results-based payments to the farmers, rewarding them based on the success of their actions [25].

Chapter 3
Enhancing water efficiency and conservation through circular agriculture solutions

Abstract: This chapter outlines strategies and innovations for enhancing water efficiency and conservation in agriculture through a circular approach. It highlights the critical role of water demand management in sustaining water resources amid climate variability and scarcity. Key strategies include precision irrigation, strategic crop selection, and the adoption of advanced technologies to optimize water use and maintain productivity. The chapter also discusses the importance of minimizing water losses through improved irrigation and nonconsumptive water practices. Moreover, it explores the integration of data analytics, artificial intelligence, and policy frameworks to support sustainable agricultural practices. This comprehensive approach combines scientific research, technological advances, and regulatory measures to ensure the long-term viability of water resources in agriculture.

3.1 Introduction

Improving water efficiency and conservation is vital in tackling water scarcity, climate change, and increasing food demands. Embracing advanced water management practices, such as precision irrigation and careful crop selection, optimizes the use of scarce water resources. These techniques minimize water wastage and bolster sustainable food production, ensuring crops are watered efficiently. Moreover, policy and regulatory support are crucial for adopting sustainable practices within the agricultural sector. This approach underscores the significance of innovation, strategic planning, and policy in advancing agricultural sustainability, securing food supplies, and preserving water resources for the future. This chapter will discuss the latest water management methodologies and technologies, including water demand management (WDM) and the benefits of precision irrigation. It will address the importance of selecting crops based on water availability and the role of advanced technologies in water use optimization. Additionally, it will cover the essential role of policy frameworks and regulatory support in enhancing water conservation efforts, highlighting how effective governance encourages adopting efficient practices. The discussion aims to provide a comprehensive understanding of achieving water sustainability in agriculture, showcasing the integration of technological innovation, strategic management, and policy support.

https://doi.org/10.1515/9783111341385-003

3.2 Understanding water demand management in agriculture

WDM prioritizes efficient water use before expanding the supply. It focuses on enhancing water efficiency across various sectors to streamline management, reduce losses, and eliminate wastage. Water demand pertains to the volumes required for irrigation, livestock, and crop cultivation in the agricultural realm. The significance of WDM in agriculture cannot be overstated – it is pivotal for protecting finite water resources, adjusting to climate variability, and realizing economic savings. This approach encompasses numerous strategies aimed at diminishing water use and boosting efficiency. Among these are the adoption of precision irrigation and the strategic selection of crops to match local water availability. An extensive range of on-farm management technologies and practices supports these efforts, enabling increased agricultural yields with reduced water consumption. These practices include the modification of crop types and the enhancement of irrigation efficiency, as well as the deployment of novel technologies for more efficient agricultural water use [26].

3.2.1 The role of water withdrawals in conservation efforts

Minimizing consumptive use, which includes water lost to evaporation or pollution, and optimizing nonconsumptive use, which allows water to be reused, are critical objectives of WDM. Addressing both is essential for reducing water losses and improving the quality of water available for reuse:

– *Consumptive use:* This refers to water that is used and not returned to its source, lost through processes like evaporation, use in crops, redirection to other basins, leakage, or contamination. These losses, which cannot be recovered, pose significant challenges to conserving water. The aim is to minimize these losses, ensuring water is used as efficiently as possible.
– *Nonconsumptive use:* This involves water that remains in its original basin after use and can be reused. However, this water often has higher levels of salts and pollutants and may require treatment to be used again effectively. Conservation efforts in this area focus on enhancing the quality of the returned water, reducing pollutant levels to make it suitable for further use, and thereby improving water efficiency [15].

3.2.2 Benefits of water conservation and efficiency

Improvements in water efficiency in agriculture aim to optimize consumptive and nonconsumptive water uses while maintaining crop production levels. To reduce overall water withdrawals, the focus is on minimizing nonbeneficial water uses, such as unnecessary evaporation and weed transpiration. This approach conserves water and generates a "new supply" that can be utilized elsewhere. For farmers, the benefits of reducing water withdrawals are significant. It lowers the cost of water acquisition and, for groundwater users, decreases the energy expenses associated with water pumping and application.

In addition, efficient water use can enhance the effectiveness of chemical applications, reducing the quantity and cost of chemicals needed. Better water management, including irrigation scheduling and efficient irrigation technologies, can improve crop quality and yields, increasing farm revenue. Furthermore, conserving water through these practices bolsters the reliability of water supplies and lessens the impact of drought and other water-related challenges. The comprehensive benefits of water conservation and efficiency in agriculture underscore the importance of strategic WDM, as detailed in Table 3.1, which highlights crucial strategies and their direct impacts on sustainability and productivity [27].

Table 3.1: Benefits of agricultural water demand management.

Demand management strategy	Benefits	Description
Runoff reduction measures	Water quality improvement	Implementing practices that reduce runoff from agricultural lands decreases contamination of water bodies with harmful substances, leading to lower treatment costs for downstream users and improved habitats for fish and wildlife
Efficient irrigation techniques	Enhanced instream flows	By optimizing irrigation, demand management conserves water, maintaining essential instream flows that support ecosystem health, including water chemistry and habitats for fish migration
Water use optimization	Protection of fish and wildlife	Strategies that minimize unnecessary water diversions help protect aquatic and terrestrial wildlife populations by preserving their natural habitats
Energy-efficient water practices	Reduced energy use	Demand management practices that reduce the total volume of water withdrawn for agriculture also decrease the energy required for water extraction and transport, saving costs and reducing greenhouse gas emissions
Soil health management	Mitigated soil salinity	Approaches that limit the volume of water and fertilizer applied to fields prevent excessive salt accumulation, protecting soil health and agricultural productivity

Table 3.1 (continued)

Demand management strategy	Benefits	Description
Infrastructure optimization	Lowered infrastructure needs	Efficient water use reduces the pressure to develop new water supply infrastructure, thus saving resources and avoiding the controversies often associated with large-scale water projects
Drought preparedness	Reduced vulnerability to drought	Demand management enhances water reserve retention, improving the reliability of water supplies and reducing agricultural vulnerability to drought conditions and future water shortages, particularly in water-stressed regions

3.3 Strategies for water efficiency

Advancing water efficiency in agriculture combines understanding crop physiology, employing agronomic practices, leveraging advanced irrigation methods, and applying engineering innovations. The focus is on selecting and managing crop types to minimize water use while ensuring productivity. Techniques like drip and sprinkler irrigation systems are essential, delivering water precisely to the plant roots and reducing wastage. Additionally, engineering improvements enhance irrigation system efficiency and bolster water storage, which is crucial for decreasing the sector's water demand. Together, these measures create a comprehensive strategy to enhance water efficiency. This approach is vital for sustainable agriculture, addressing water scarcity and climate change impacts, and safeguarding water resources while maintaining economic viability.

3.3.1 Crop types and physiology

Addressing water scarcity in agriculture requires a nuanced understanding of crop types and physiology, which is pivotal for advancing water efficiency. Strategic WDM becomes crucial with climate change, exacerbating drought conditions globally. This strategy encompasses several vital components.

Understanding the diverse water needs and responses of different crops to water stress is fundamental. Crop physiology offers insights into how plants utilize water, their efficiency in converting water into biomass, and their resilience to water shortages. This knowledge supports the strategic selection of crop types, aligning with local climatic conditions and water availability to optimize usage and maintain productivity.

Key strategies include:
- *Strategic crop selection*: Choose drought-resistant varieties that require less water by leveraging genetic traits such as deeper root systems for better water access and efficient leaf structures to minimize transpiration.
- *Crop phenology*: Water must be applied judiciously at critical growth phases to ensure crops receive adequate moisture without wasteful irrigation.
- *Agronomic techniques*: Implement mulching to reduce soil moisture evaporation and soil management to enhance water infiltration and retention, lowering the overall water demand.

Furthermore, integrating crop diversification into farm management can spread risk and improve resilience to climatic variability. By cultivating a mix of crop types with different water needs and resistance to water stress, farmers can create more stable production systems that are less susceptible to water shortages or droughts.

Finally, advancements in irrigation techniques and adopting efficient water management technologies play a crucial role. Precision irrigation targets water delivery directly to the plant roots, and smart irrigation systems informed by real-time data on soil moisture and weather conditions can significantly enhance water use efficiency [15, 28–30].

3.3.2 Agronomic techniques

Agronomic techniques are essential for achieving water efficiency in agriculture, offering practical solutions to the challenges of water scarcity while ensuring sustainable crop production. These techniques encompass a broad range of practices designed to optimize water use at every stage of the agricultural process, from soil preparation to crop cultivation. There are a range of practices that have an impact on water conservation:
- *Soil moisture conservation*: Techniques such as cover cropping and mulching play a pivotal role in preserving soil moisture. By reducing evaporation and improving water retention, these practices decrease the frequency of irrigation needed, contributing significantly to water savings.
- *Optimized irrigation scheduling*: Adapting irrigation schedules to the specific water needs of crops, considering their various growth stages, minimizes excess water use. Employing soil moisture sensors or adopting evapotranspiration (ET)-based scheduling can pinpoint the most effective timing and quantities for irrigation.
- *Drought-resistant cropping systems*: Using crops or varieties resistant to drought conditions enables agriculture to maintain productivity even in water-limited environments. These plants are genetically adapted to thrive with less water, making them invaluable in water-stressed regions.

- *Efficient nutrient management*: Proper nutrient management, especially for nitrogen and phosphorus, can enhance a crop's water efficiency. Techniques that reduce nutrient runoff and promote effective uptake support healthier growth and reduce the water needed for crop success.
- *Crop rotation and intercropping*: These practices improve soil health, improving water infiltration and retention. They also disrupt pest and disease cycles, potentially reducing the need for chemical treatments that might demand additional water for application.
- *Tillage practices*: Conservation tillage methods, including minimal or no tillage, help maintain soil structure, reduce erosion, and keep organic matter intact. This results in soil absorbing and retaining water more effectively, aiding drought resistance.
- *Water harvesting and recycling*: Capturing and reusing water from rainfall or agricultural runoff can provide an additional water source for irrigation. On-farm storage solutions, such as ponds or reservoirs, allow for the collection and use of water that would otherwise be lost, offering a buffer against dry periods [15, 31–36].

3.3.3 Foundations of field irrigation techniques

Efficient water management in agriculture increasingly depends on adopting advanced field irrigation techniques. These methods significantly reduce water loss while ensuring crops receive the precise water they need for optimal growth. By moving away from traditional flood irrigation practices toward more targeted irrigation systems, such as drip and sprinkler systems, farmers can enhance water delivery directly to the plant roots, where it is most needed. This approach conserves water and supports healthier and more productive crops. Advanced field irrigation techniques include:

- *Drip irrigation*: This system delivers water directly to the base of each plant through a network of tubing and emitters. Drip irrigation maximizes water efficiency by minimizing evaporation and runoff, making it ideal for row crops and perennial plantations. The precise delivery also reduces weed growth and limits the potential for waterborne diseases.
- *Sprinkler systems*: Sprinklers can provide more uniform water distribution over a larger area and are adaptable to various terrain and soil conditions. When managed correctly, sprinkler irrigation can reduce water usage compared to traditional methods, especially when equipped with timers and moisture sensors for optimized scheduling.
- *Irrigation scheduling*: Modern irrigation strategies are increasingly informed by a detailed understanding of crop water requirements and soil moisture levels. Technologies such as soil moisture sensors, weather data analysis, and crop ET modeling inform irrigation schedules, ensuring water is applied only when neces-

sary and in the amounts required. This conserves water and prevents over-irrigation, which can lead to nutrient leaching and reduced crop quality.
- *Automation and control systems*: Integrating automation into irrigation systems allows for more precise control over water application. Automated systems can adjust watering based on real-time data, reducing labor costs and enhancing water use efficiency. Remote monitoring and control via smartphones or computers further simplify management, allowing adjustments to be made from anywhere at any time.
- *Alternative water sources*: Exploring nonconventional water sources for irrigation, such as treated wastewater or collected rainwater, can supplement traditional water supplies. This approach is valuable in regions experiencing water scarcity, providing a sustainable alternative that reduces the demand for freshwater resources [13, 15].

3.3.4 Foundation for efficient water management

Foundational engineering solutions are essential in sustainable agriculture to tackle water scarcity and improve irrigation efficiency. This section outlines key water management strategies that form the basis for efficient agricultural practices, including:
- *Water recycling systems*: Water recycling technologies are central to reducing freshwater withdrawals. These systems capture runoff or wastewater from agricultural processes, treat it, and then reuse it for irrigation. This conserves water and minimizes the agricultural sector's impact on local water bodies by reducing nutrient and chemical runoff.
- *Advanced irrigation infrastructure*: Upgrading traditional irrigation infrastructure to incorporate advanced materials and designs can significantly reduce water loss. For instance, using nonporous materials in canal and pipeline construction minimizes seepage, while precision-engineered components in irrigation systems reduce leaks.
- *Automated irrigation controls*: Engineering solutions now often include integrating automation and sensor technology into irrigation systems. These controls can adjust water flow based on real-time soil moisture data and weather conditions, ensuring that water is applied efficiently and only as needed.
- *Evaporation minimization*: Techniques to minimize evaporation from open water surfaces include covering reservoirs and applying chemical monolayers that reduce evaporation rates. These methods are particularly effective in hot, dry regions where evaporation significantly increases water loss.
- *Leak detection technology*: Advanced leak detection systems, utilizing acoustic, thermal, or chemical tracing methods, enable early identification and repair of leaks within irrigation networks. This conserves water and ensures that the maximum volume reaches the intended crops.

–   *Water-efficient conveyance systems*: Another crucial engineering approach is re-designing conveyance systems to minimize water loss during transportation. This may involve transitioning from open channels to closed piping systems or lining canals with impermeable materials to prevent seepage.
–   *Use of nontraditional water sources*: Engineering techniques also facilitate using alternative water sources for irrigation, such as desalinated water or treated gray water. The infrastructure capable of handling these water types can expand water resources available for agriculture without increasing stress on conventional water sources [13, 15].

Case 3.1 Enhancing forage production and climate resilience through crop diversification in Slovenia

In Slovenia, a project aims to enhance forage production in drought and climate change by implementing crop rotation and diversification strategies. The initiative focuses on six test farms to compare the effects of using winter legume catch crops, both alone and mixed with grasses, against the cultivation of lucerne (*Medicago sativa*) for its drought resistance and potential to improve animal feed quality and quantity. Lucerne, known for stable yields and suitability for conservation, is evaluated for its feeding value and fermentation quality. The project highlights the benefits of integrating more legumes into crop rotations and capturing more nitrogen for fertilization, thereby improving soil fertility and reducing the reliance on mineral nitrogen fertilizers. This approach maintains the arable land's potential for food production and enhances it due to the positive effects on soil health and subsequent crops. Despite the traditional use of Italian ryegrass among Slovenian farmers, there is growing interest in producing forage with lucerne. The project aims to present its findings as innovative practices to farmers. It demonstrates how these methods can yield high-quality winter forage, improve soil health, reduce dependency on external resources, and increase farm resilience against climate change challenges [37].

3.4 Advanced technologies in water efficiency

Adopting advanced technologies significantly enhances water efficiency in agriculture, tackling the dual challenges of water scarcity and the need for precise irrigation. Innovations like precision irrigation and automated systems optimize water delivery, reducing waste and supporting crop growth. These technologies enable targeted watering, customizable application rates, and soil moisture monitoring, dramatically increasing efficiency. Engineering solutions improve water distribution and storage, while information and communication technology (ICT) tailors irrigation practices to real-time conditions. Smart management of ET further ensures that water use aligns with crop requirements, underpinning a sustainable approach to agriculture amid growing environmental pressures.

3.4.1 Advanced irrigation technologies: precision and automation

Enhancing field irrigation techniques by integrating cutting-edge technologies and advanced systems is pivotal for optimizing water use in agriculture. These modern approaches focus on delivering water precisely where and when needed, drastically reducing wastage and ensuring crops receive adequate hydration for optimal growth. This section delves into two critical advancements in field irrigation techniques: precision irrigation and application of automation and control systems.

3.4.1.1 Precision irrigation
Precision irrigation techniques transform how water is applied in agriculture, leveraging advanced technologies to optimize irrigation and ensure water reaches where it is most needed. These techniques include:
- *Targeted water delivery*: Precision irrigation systems, such as drip and micro-sprinkler systems, deliver water directly to the crop's root zone. This method significantly reduces losses from evaporation and runoff, ensuring water is utilized where it is most beneficial.
- *Variable rate irrigation (VRI)*: VRI technology allows for the customization of water application rates across different field sections, accommodating variations in soil types, crop stages, and moisture levels. This adaptability ensures that each part of the field receives precisely the water it needs, maximizing efficiency.
- *Soil moisture monitoring*: In precision irrigation, soil moisture sensors are used. These devices provide real-time data on soil water content, allowing for the precise scheduling of irrigation events to avoid over- or under-watering [38–40].

3.4.1.2 Automation and control systems
Automation and control systems bring a new efficiency level to agricultural irrigation, enabling precise management and adaptability. These systems encompass:
- *Automated irrigation scheduling*: Leveraging data from soil moisture sensors, weather stations, and ET models, automated irrigation systems can decide the optimal irrigation times. This conserves water and can lead to significant energy savings by reducing the need for pumping during peak hours.
- *Remote management*: Modern control systems can adjust irrigation settings remotely via smartphones or computers. This flexibility allows immediate responses to weather conditions or soil moisture levels, enhancing water use efficiency.

- *Integration with crop management systems*: Automation technologies are increasingly integrated with broader crop management systems. This holistic approach allows for the coordination of irrigation with other farming activities, such as fertilization and pest management, further optimizing resource use [41, 42].

3.4.1.3 Remote sensing and GIS for water management

Remote sensing (RS) and geographical information systems (GISs) are increasingly central to sustainable agriculture, particularly water resource management. These technologies provide essential data on water availability, crop health, soil moisture, and environmental shifts, supporting targeted and efficient water management practices.

RS technologies utilize satellite and aerial imagery to survey the Earth's surface. They supply vital information on the size and condition of water bodies, the impact of water scarcity on crops, and land use changes that influence water resources. By analyzing RS data, including the Normalized Difference Vegetation Index from satellite images, experts can gauge plant health and respond to water stress. This information enables timely irrigation and water distribution adjustment, enhancing water efficiency and crop resilience.

GIS are instrumental in managing spatial data related to water resources. GIS facilitates the creation of detailed maps incorporating data on terrain, soil types, water sources, and usage, aiding in identifying areas for water conservation, irrigation planning, and risk assessment for droughts or floods. Through GIS, models can predict water flow and distribution, supporting strategic water management decisions [14, 43, 44].

3.4.1.4 Data analytics and water usage

Integrating data analytics into agricultural water management signifies a significant shift toward sustainability. Utilizing big data enables the transition from conventional practices to more precise, resource-conserving strategies that improve productivity:

- *Data analytics and water usage*: Data analytics enhances water use efficiency in modern agriculture. By analyzing extensive datasets that include soil moisture, crop water consumption, weather patterns, and irrigation performance, data analytics offers detailed insights into water requirements and how best to fulfill them.
- *Big data in water management*: The use of big data involves gathering and analyzing varied data sources such as satellite imagery, sensor networks, weather stations, and irrigation systems. This extensive data collection supports in-depth water demand analysis for various crops under different conditions, promoting the adoption of precision irrigation methods. These methods not only minimize wastewater but also maximize yields. These vast datasets are turned into actionable insights through big data technologies, helping to efficiently tailor irrigation schedules and water distribution to meet specific crop needs and environmental conditions.

– *Predictive analytics for water demand forecasting*: At the leading edge of data-driven water management is predictive analytics, which uses machine learning and statistical models to predict future water needs and supply. By evaluating historical weather, crop water use, and soil moisture data, predictive models forecast upcoming water requirements, enhancing water allocation efficiency. This proactive strategy is vital for adapting to climate change, managing resources in drought conditions, and maintaining water supply sustainability [45–47].

3.4.1.5 Artificial intelligence in irrigation management

Integrating artificial intelligence (AI) into irrigation system management represents a significant advancement in agricultural technology, enhancing water efficiency and crop productivity. AI applications in irrigation leverage predictive analytics, machine learning models, and automated decision-making to optimize water use tailored to the precise needs of crops at various growth stages. AI in irrigation management involves:

– *Predictive analytics*: AI utilizes historical and real-time data to forecast future water requirements, climate conditions, and potential crop stressors. This anticipatory approach allows for planning irrigation schedules that align closely with anticipated conditions, ensuring water is used judiciously and effectively.

– *Machine learning models*: By analyzing data from sensors, weather stations, and other sources, AI develops models that improve over time, identifying the most efficient irrigation methods for various scenarios. This continuous learning process enhances decision-making, improving water conservation and crop yields.

– *Automated decision-making*: AI algorithms can automate irrigation decisions, determining the optimal timing and volume of water delivery. This reduces manual intervention, ensuring a consistent and scientifically grounded approach to irrigation that adapts to changing conditions.

– *Integration with Internet of things (IoT) devices*: With the IoT technology, AI facilitates a highly responsive irrigation system. Sensors provide real-time data on soil moisture and environmental factors, which AI processes to make immediate irrigation adjustments [46, 48].

Case 3.2 Cloud-based water management system for improving irrigation efficiency and sustainability in Australia

The COALA project is a Copernicus-based information service that supports Australian farmers in the Murray-Darling Basin by using satellite data for precision irrigation and nutrient management. The COALA project partnered with Rubicon Water to develop a cloud-based water management system to improve irrigation efficiency and reduce water usage in the Murray Darling Basin, an Australian agricultural region. The system involved installing sensors and monitoring devices throughout the irrigation network to provide real-time data on water usage, weather patterns, and other factors impacting irrigation efficiency. The data was analyzed using the COALA project's logistics analytics solutions to identify optimization opportunities, predict future demand, and improve resource allocation. The project resulted in a 20% improvement in irrigation efficiency, significant cost savings for farmers, and a reduced environmental impact. In addition, the project demonstrates the potential for cloud-based logistics analytics solutions to be applied to other agricultural areas worldwide to improve efficiency and sustainability [46].

3.4.2 Engineering techniques for optimizing water distribution and storage

Advanced engineering solutions are reshaping water distribution and storage in agriculture, significantly enhancing efficiency. These innovations are crucial in lowering water use and improving farm resilience against variable water availability. The following delves into cutting-edge developments such as advanced leak detection, adaptive smart irrigation, and extensive rainwater harvesting (RWH), aiming for the sustainable optimization of water resources.

3.4.2.1 Efficient water distribution systems
Efficient water distribution systems are crucial for maximizing agricultural productivity and minimizing waste. Key components include:
– *Leak detection and repair technologies*: The integration of advanced leak detection systems, utilizing acoustic, thermal imaging, or satellite technology, enables the early identification and repair of leaks within irrigation infrastructure. This proactive approach prevents water loss, ensuring a more significant water volume reaches the crops.
– *Irrigation canal upgrades*: Upgrading existing irrigation canals with liners or covers minimizes water loss due to seepage and evaporation. Implementing precision-engineered materials and designs enhances water conveyance efficiency, delivering water to fields with minimal waste.
– *Smart irrigation controls*: Adopting smart valves and flow meters controlled by sophisticated software allows for dynamic adjustments in water delivery based on real-time demand and conditions. This system-level optimization of irrigation practices reduces unnecessary water use and improves overall water management on the farm [15, 49–51].

3.4.2.2 Innovative water storage techniques
Innovative water storage techniques enhance agricultural resilience by ensuring water availability during critical periods. These techniques involve:
– *On-farm reservoirs*: Developing more efficient on-farm reservoirs equipped with impermeable liners or covers significantly reduces evaporation losses. These res-

ervoirs serve as critical assets in capturing and storing rainwater or runoff for later use, particularly during dry spells.

–   *Underground storage solutions*: Underground storage techniques, such as aquifer recharge or the construction of underground reservoirs, provide a method to bank excess water during periods of abundance. This stored water can be a vital resource during drought conditions, ensuring continuous water availability for irrigation [52].

3.4.2.3 Rainwater harvesting systems

Implementing RWH systems involves capturing and storing water from surfaces such as greenhouse roofs or farm buildings. This approach supplements irrigation supplies, reducing dependency on external water sources and maximizing the use of all available water. RWH systems can be relatively simple or more complex, depending on the size and needs of the agricultural operation.

Collected rainwater can be stored in tanks and used during dry periods, providing a reliable water source when traditional supplies are limited. This enhances resilience against water shortages and helps maintain consistent crop growth and productivity. Farmers can significantly reduce their water costs by utilizing harvested rainwater, making their operations more economically sustainable.

In addition to financial benefits, RWH contributes to environmental conservation. It reduces the strain on local water supplies, which is particularly important in areas experiencing water scarcity. This method also mitigates the impact of stormwater runoff, which can cause soil erosion and water pollution. By capturing and using rainwater, agricultural operations can lower their environmental footprint and promote more sustainable farming practices.

There are several types of RWH systems, including surface runoff harvesting and rooftop harvesting. Surface runoff harvesting involves collecting rainwater from open spaces or fields and directing it to storage structures like ponds or underground tanks. Rooftop harvesting captures rainwater from roofs and channels it through gutters and downspouts into storage tanks or cisterns.

Several factors must be considered to implement an effective RWH system, including the collection surface area, storage capacity, and filtration requirements. Proper system maintenance is essential to ensure the quality of the harvested water and the longevity of the equipment [14, 15].

Case 3.3 Enhancing agricultural sustainability through RWH in India

RWH and artificial recharge of groundwater (ARG) are vital for sustainable farming in India, where agriculture largely depends on monsoon rains. By capturing and storing rainwater, these practices mitigate the impacts of water scarcity, ensuring a reliable water supply for irrigation and enhancing groundwater levels. For instance, implementing RWH systems has enabled farmers to improve crop yields significantly. A study in Andhra Pradesh reported that RWH led to a cropping intensity increase of 152% from a baseline of 96%, demonstrating the profound impact on agricultural productivity. Furthermore, the government's support through schemes like the Mahatma Gandhi National Rural Employment Guarantee Scheme has facilitated the construction of over 293,873 water conservation structures, showcasing a nationwide commitment to enhancing water security for agriculture. These initiatives support the livelihood of millions of farmers and contribute to the agricultural sector's overall sustainability, making RWH and ARG essential components of India's strategy to combat climate change and water scarcity [53].

3.4.3 Management of irrigation systems

The management of irrigation systems is transforming by integrating ICT tools and advanced analytical systems. Various technologies are being applied to revolutionize agricultural water use, facilitating more precise and efficient irrigation practices. By harnessing ICT and decision support systems (DSSs) and RS technologies, farmers and water managers can achieve optimal water distribution and utilization tailored to crops' specific needs at various growth stages.

3.4.3.1 Integration of ICT tools in irrigation management
The integration of ICT tools in irrigation management brings precision and efficiency to water use in agriculture, characterized by:
- Real-time monitoring and control: The adoption of ICT tools allows for the real-time monitoring of irrigation systems, enabling immediate adjustments to water distribution based on current field conditions. This can include changes in soil moisture levels, weather patterns, and crop water needs, ensuring that water is applied efficiently and only as necessary.
- Data analytics for water optimization: ICT systems collect vast amounts of data from sensors distributed throughout the field. By analyzing this data, farmers can gain insights into the optimal timing and quantity of irrigation, reducing water use while maintaining or enhancing crop yields [54, 55].

3.4.3.2 Decision support systems and remote sensing
DSS and RS technologies refine water management in agriculture through:
- Precision irrigation scheduling: DSSs integrate weather forecasts, soil moisture data, and crop ET rates to generate precise irrigation schedules. These systems help determine the most effective irrigation timings and quantities, improving water use efficiency.

- Crop water requirement calculations: Advanced DSS algorithms calculate different crops' exact water requirements at various growth stages. This ensures that crops receive adequate hydration without the risk of over-irrigation, which can lead to water wastage and nutrient leaching.
- Early detection of plant stress: RS technologies, including satellite imagery and drone-based sensors, can identify signs of plant stress early on. This enables timely interventions to address water deficits before they impact crop health and productivity.

Case 3.4 Weenat's innovative approach to water management through ICT and AI integration

Weenat's innovative approach to agricultural water management integrates a Europe-wide network of over 20,000 soil sensors with satellite imagery and AI to provide real-time soil moisture data. This project, aimed at addressing climate change impacts and optimizing water use, has been recognized by the i-Nov competition for its potential to enhance the efficiency of water resources management. Agriculture, which accounts for 70% of global freshwater withdrawals, faces significant challenges from climate-induced water scarcity. Weenat's technology targets this issue by enabling precise water management and reducing reliance on extensive water withdrawals. In 2023 alone, Weenat's sensors saved 32 million m^3 of water, equivalent to approximately 12,000 Olympic-sized swimming pools. This substantial water conservation effort underpins the project's goal to support nearly 1 million European irrigators, underscoring the critical integration of ICT tools in achieving sustainable and efficient agricultural practices [56].

3.4.4 Evapotranspiration management

Efficient management of ET within agricultural systems is crucial for optimizing water use and enhancing irrigation efficiency. Various advanced techniques and technologies aim to manage ET more effectively, ensuring that crops receive the precise amount of water they need for optimal growth while minimizing waste, including the following.

3.4.4.1 Advanced techniques for ET management

Advanced techniques for ET management enhance irrigation precision, utilizing:

- Crop coefficients and satellite data: By applying crop coefficients – a factor that reflects the water needs of a specific crop at various stages of growth – in conjunction with satellite data, farmers can accurately estimate ET rates across different fields. This precise calculation allows for adjusting irrigation schedules and volumes to match the actual water requirements of crops, significantly improving water use efficiency.

- Satellite imagery for ET estimation: Satellite imagery provides a broader perspective, offering insights into field-level ET rates. When analyzed alongside local weather conditions, this data can help create a more accurate and comprehensive understanding of water loss through ET across more significant agricultural landscapes.

3.4.4.2 Integration of smart technologies

Integration of smart technologies streamlines water management, featuring:

- Smart sensors for real-time monitoring: The adoption of smart sensors in fields enables continuous monitoring of ET and soil moisture levels. These sensors provide data that can be used to make immediate adjustments to irrigation practices, ensuring that water is applied in the right amounts at the right time.
- IoT-based solutions: IoT technology connects sensors and irrigation equipment to a central management system. This setup allows for the automated adjustment of irrigation based on real-time ET and soil moisture data, streamlining the irrigation process and reducing the need for manual intervention [57–60].

Case 3.5 Advancing evapotranspiration management in Australia

The Soil Moisture Integration and Prediction System (SMIPS), developed by the Terrestrial Ecosystem Research Network in Australia, represents a significant advancement in ET management within agricultural systems. Utilizing a data model fusion approach, SMIPS delivers daily, nationwide estimates of volumetric soil moisture at approximately 1 km resolution. This innovative system combines ground-based soil moisture data, satellite imagery, and climatic data from various sources, including the Bureau of Meteorology and the European Space Agency's Soil Moisture and Ocean Salinity mission, to produce accurate soil moisture readings and an index indicating the moisture content in the top 90 cm of soil. By providing such precise and accessible soil moisture information, SMIPS enables farmers and land managers to optimize irrigation schedules and water volumes to match the actual requirements of crops, significantly improving water use efficiency and contributing to sustainable agricultural practices. This system addresses the critical need for managing ET more effectively, thereby minimizing water wastage and enhancing irrigation precision [61].

3.5 Policy and regulatory frameworks

Implementing policy and regulatory frameworks is crucial for sustainable water use in agriculture, addressing water scarcity and ensuring equitable resource distribution. Water rights and allocation policies establish legal frameworks to allocate water efficiently among users, protecting the environment and user rights. Incentives for adopting water-efficient technologies, such as subsidies and tax breaks, encourage farmers to invest in sustainable practices. Regulations on water quality aim to mitigate agricultural runoff, setting standards for chemical use and managing pollutants. Public awareness and education programs also play a crucial role in promoting water conservation, providing farmers and the community with the knowledge to implement efficient water use practices. Together, these measures form a comprehensive approach to managing agricultural water sustainably, safeguarding resources for future generations.

3.5.1 Implementation of water rights and allocation policies

Implementing water rights and allocation policies is a pivotal strategy for managing agricultural water use in an increasing water scarcity and resource competition era. Such policies are instrumental in establishing a structured and equitable system for water distribution, ensuring that water is allocated efficiently among agricultural users. This approach prioritizes the optimal use of water resources, maximizing value while safeguarding the rights of all users and the environment.

Components of regulatory frameworks on water quality and use are essential for ensuring sustainable agricultural practices and safeguarding water resources. These components include:

–   Establishment of water rights systems: Water rights systems formalize water entitlement to various users, including farmers, based on legal or customary rights. These systems are crucial for clarifying who can use water, how much they can use, and under what conditions, providing a clear framework for water allocation.
–   Efficient allocation among users: The core objective of water allocation policies is to distribute water in a manner that supports the most valuable and sustainable uses. By doing so, policies aim to optimize the economic benefits of water use within the agricultural sector and beyond, ensuring that water contributes to overall societal well-being.
–   Protection of user rights: Effective water rights and allocation policies protect the interests of all users, including smallholder farmers and indigenous communities, by ensuring fair access to water resources. This is essential for maintaining the livelihoods of those who depend on water for agriculture and preventing conflicts over water use.
–   Environmental safeguards: Beyond human uses, these policies incorporate provisions to maintain environmental flows necessary to support aquatic ecosystems

and the services they provide. By setting aside water for ecological needs, policies help maintain water bodies' health, which is vital for long-term water security.

–   Enforcement mechanisms: The success of water rights and allocation policies hinges on robust enforcement mechanisms. This includes monitoring water use, penalizing unauthorized withdrawals, and resolving disputes over water rights fairly and promptly.

–   Adaptability and flexibility: Given the variability of water availability due to seasonal changes and climate variability, policies must be adaptable. Flexibility allows for adjusting water allocations in response to droughts, floods, and long-term changes in water supply and demand patterns [14, 15].

Case 3.6 Water trading in the Murray–Darling Basin

Water trading in the Murray–Darling Basin allows users to buy and sell water rights, promoting efficient water use and supporting agricultural productivity. There are two main types of water trades: permanent trade, where water entitlements are sold permanently, and temporary trade, where annual water allocations are traded based on seasonal needs. Water entitlements grant holders a long-term right to a portion of the available water in a system. In contrast, water allocations are the amount of water distributed each year, influenced by rainfall, inflows, and stored water levels. The ability to trade water provides irrigators with flexibility, allowing them to adjust water use according to their business requirements, crop cycles, and climate conditions. For example, in wet years, entitlement holders may need less water and can sell their surplus, while in dry years, they may buy additional water to meet demand. This market-driven system ensures water can be allocated to its highest-value uses, promoting more productive and sustainable agriculture. Water trading is a vital economic tool in the Murray–Darling Basin, with the market valued at approximately $4 billion annually. Most water traded is surface water, though some groundwater is also traded. The Murray–Darling Basin Authority oversees the fairness and transparency of water markets, while state governments in New South Wales, Queensland, South Australia, and Victoria manage the allocation processes. This system provides farmers with greater choice, reduces financial risk, and contributes to sustainable water management across one of Australia's most important agricultural regions [62].

3.5.2 Incentives for water-efficient technologies

Incentivizing the adoption of water-efficient technologies and practices in agriculture is a strategic approach to enhancing water conservation and efficiency. Through targeted policies, governments and water management authorities can motivate farmers to integrate innovative solutions that significantly reduce water use while maintaining or even improving crop yields. These incentives can take various forms, including subsidies, tax breaks, and grants, making it economically viable for farmers to invest in water-saving technologies. Forms of incentives for water-efficient technologies include the following:

– Subsidies for efficient irrigation systems: Subsidies can lower the financial barriers to acquiring advanced irrigation equipment, such as drip or sprinkler systems, which are known for their efficiency in water use. By partially or fully covering the cost of these systems, subsidies make it more accessible for farmers to upgrade their irrigation practices.

– Tax breaks for water conservation investments: Offering tax incentives for investments in water-saving technologies encourages farmers to integrate these practices into their operations. Tax breaks can provide significant savings for farmers, making it more appealing to invest in technologies that might otherwise be considered too costly.

– Grants for implementing water conservation measures: Grants provided to farmers to implement specific water conservation measures, such as soil moisture monitoring systems or water recycling facilities, can support adopting comprehensive water management strategies on farms [14, 15].

Case 3.7 Washington's Irrigation Efficiencies Grant Program

The Irrigation Efficiencies Grant Program (IEGP), established in 2002, provides financial incentives to Washington's agricultural irrigators and water purveyors to install more efficient irrigation systems. Administered by the State Conservation Commission in partnership with conservation districts, the program aims to reduce water demand, enhance water quality, and restore instream flows, benefiting agricultural productivity and ecosystems. The program was initially focused on improving irrigation efficiency in 16 Salmon Critical Basins to protect water flows for salmonids listed as endangered or threatened under the Endangered Species Act. Adaptive management expanded the program's scope to include other water-scarce basins. Water rights saved through irrigation upgrades are transferred to Washington's Trust Water Rights Program to ensure that the conserved water remains instream for environmental purposes. Grants from the IEGP fund technicians who assist irrigators by evaluating water-saving projects, designing new systems, and developing irrigation water management plans. These upgrades include improving conveyance systems like ditches and enhancing application systems like flood or sprinkler irrigation. Technicians also monitor project outcomes to ensure ongoing water savings and system efficiency. Eligible projects must address drought vulnerability, water quality protection, and instream flow enhancement. In some cases, saved water may need to be transferred to the state's Trust Water Rights Program to protect the project's public benefits [63].

3.5.3 Regulations on water quality and use

Implementing water quality and use regulations is a critical approach to mitigating the impacts of agricultural activities on water resources. By establishing regulatory frameworks, authorities can set enforceable standards that prevent the degradation of water bodies from agricultural runoff, which is often laden with fertilizers, pesticides, and other pollutants. These regulations are designed to protect water quality and ensure that water use within agriculture is sustainable and responsible.

The components of regulatory frameworks on water quality and use are crucial for promoting sustainable agricultural practices and protecting water resources. Key aspects include:

– Limits on fertilizer and pesticide use: Regulations can specify allowable types and quantities of fertilizers and pesticides, minimizing the risk of runoff that leads to nutrient pollution and chemical contamination of water bodies. These limits encourage adopting best management practices, such as precision application and integrated pest management, to reduce dependency on chemical inputs.
– Runoff management practices: Mandating the implementation of runoff management practices is critical to controlling the flow of pollutants into waterways. This may include the construction of buffer zones, sediment traps, and wetlands designed to filter and capture runoff before it reaches rivers and lakes.
– Water use permits: Regulatory frameworks often require agricultural operations to obtain permits for water withdrawals, setting clear boundaries on the volume of water that can be used. This ensures that water use is monitored and managed to prevent overextraction, which can deplete water sources and harm aquatic ecosystems.
– Nutrient management plans: Regulations may necessitate developing and adhering to nutrient management plans that outline how fertilizers should be applied to minimize leaching and runoff. These plans are tailored to specific crops and soil types, optimizing nutrient use efficiency and protecting water quality [14, 15].

Case 3.8 California's agricultural water efficiency and management requirements

California's robust agricultural success, marked by its role as the sole producer of 13 commodities and a top producer of over 74 commodities in the United States, is significantly supported by strategic water management. The state's agricultural sector, drawing upon approximately 34 million acre-feet of water annually to irrigate 9.6 million acres, demonstrates high water use efficiency. This efficiency is essential, given that agriculture accounts for about 40% of California's water use. The Water Conservation Act of 2009 mandates Agricultural Water Management Plans for suppliers serving over 25,000 irrigated acres to enhance this efficiency. These plans emphasize the adoption of efficient water management practices, focusing on optimizing water distribution and usage. Through such measures, California aims to ensure that water resources are judiciously used, bolstering the agricultural sector's resilience against water scarcity while supporting the state's environmental and economic goals. This initiative reflects a concerted effort to balance agricultural productivity with sustainable water resource management [64].

3.5.4 Public awareness and education programs

Public awareness and education programs are crucial to water conservation policies, designed to enhance understanding and adoption of water-efficient practices among farmers and the broader community. These initiatives aim to foster a culture of water stewardship, emphasizing the critical importance of conserving water resources for the sustainability of agriculture and the well-being of ecosystems. By providing targeted educational programs and resources, these policies empower individuals and communities to make informed decisions and adopt behaviors that contribute to more efficient water use.

Critical elements of public awareness and education programs are vital for fostering a culture of sustainable water management in agriculture and beyond. These elements include:

- Targeted educational campaigns: Launch campaigns highlighting the significance of water conservation in agriculture and its impact on food security and environmental health. These campaigns can use various media platforms to reach a broad audience and effectively communicate the need for sustainable water management practices.
- Workshops and training for farmers: Organize workshops and training sessions that provide farmers with practical knowledge on implementing water-efficient technologies and practices. Topics might include precision irrigation, soil moisture management, and drought-resistant crop varieties, equipping farmers with the skills to reduce water use while maintaining crop productivity.
- Educational materials and resources: Develop and distribute educational materials, such as brochures, videos, and online content, that offer guidance on water-saving techniques and best agriculture practices. These resources can serve as valuable tools for farmers and the public to learn how to conserve water in their operations and daily lives.
- School and community programs: Integrating water conservation education into school curriculums and community programs to instill an understanding of water issues from an early age. These programs can encourage students and community members to participate in water conservation projects and initiatives, fostering a collective effort toward sustainable water use.

–   Collaboration with agricultural extension services: Partnering with agricultural extension services to directly disseminate information on water conservation and efficient water use practices to farmers. Extension agents can play a crucial role in providing personalized advice and support, helping to translate knowledge into actionable changes on the ground [14, 15].

Case 3.9 North Carolina State University's Soil & Water Resource Conservation Workshop

The Soil & Water Resource Conservation Workshop is a weeklong educational program held at North Carolina State University, designed for high school students interested in natural resource conservation. Students nominated by their local soil and water conservation district participate in hands-on activities and studies on environmental conservation, resource management, and sustainability. The program gives priority to rising juniors and seniors. The workshop provides an immersive learning experience, introducing students to conservation topics such as soil health, water management, and environmental challenges in a global context. In addition to educational activities, students can win awards and scholarships. For example, the Conservation District Employees Association Award offers a $1,000 college scholarship to a student pursuing higher education in natural resources or environmental science. Other awards, including the S. Grady Lane Award and additional academic performance-based scholarships, recognize students who demonstrate excellence in cooperation, behavior, and academic achievement during the workshop. Participants engage with natural resource professionals, gaining valuable insights and guidance. This workshop not only fosters environmental awareness but also encourages students to pursue careers in conservation [65].

Chapter 4
Sustainable agriculture and nature-based solutions preserving water quality

Abstract: This chapter examines the impact of conventional agriculture on water quality, focusing on the challenges of chemical use, soil erosion, and nutrient runoff. To counter these effects, it advocates for sustainable and nature-based solutions (NBSs) like constructed wetlands, riparian buffers, agroforestry, and biofiltration. It highlights successful NBS implementation case studies and illustrates how such practices can improve water quality, agricultural productivity, and ecosystem health. The discussion emphasizes the importance of integrating eco-friendly strategies into agriculture to support sustainability, showcasing the potential of sustainable agriculture and NBS in enhancing environmental quality and resilience.

4.1 Introduction

In recent years, agricultural practices have significantly impacted water quality, surpassing the effects of industrial and urban sources in many regions. This increase in pollution is mainly due to the intensive use of fertilizers, pesticides, manure runoff, and soil erosion, which contribute to high levels of nitrates, phosphorus, and sediments in water bodies. Addressing this issue requires an integrated approach that balances ecosystem health with agricultural productivity. This chapter explores sustainable agriculture and nature-based solutions (NBSs) within the circular economy framework as critical strategies for reducing farming's impact on water quality. It discusses the challenges of traditional farming, the benefits of sustainable practices, and the role of nature-based methods in enhancing water quality and ecosystem health [66, 67].

4.2 Impact of conventional agriculture on water quality

Conventional agriculture significantly impacts water quality through chemical fertilizers, pesticides, soil erosion, and nutrient runoff. This approach leads to water pollution and affects aquatic ecosystems, human health, and biodiversity.

4.2.1 Chemical fertilizers and pesticides

Using chemical fertilizers and pesticides in agriculture significantly affects water quality and aquatic life, posing a complex environmental challenge. Fertilizers, rich

https://doi.org/10.1515/9783111341385-004

in nutrients like nitrates and phosphorus, enhance crop growth but can cause the over-enrichment of water bodies, leading to eutrophication. This process promotes algal blooms that, when decomposed, consume large amounts of oxygen, resulting in low oxygen levels (anoxia) that can severely harm fish and other aquatic organisms, potentially causing mass die-offs. Pesticides, intended to control pests and weeds, often contain compounds toxic to nontarget species, including fish and invertebrates critical to the aquatic food web.

Chemical contaminants from agricultural runoff enter waterways through direct application, spray drift, and surface runoff, impacting water quality far from their sources. This runoff can lead to local losses of fish and invertebrates, with certain pesticides like synthetic pyrethroids being particularly harmful. Aquatic ecosystems depend on a delicate balance of species, and the loss of even one species can disrupt the entire system. For instance, losing invertebrates may force fish to travel farther for food, increasing their exposure to predators and other dangers.

These contaminants also increase turbidity and decrease water clarity by stimulating the growth of algae and microorganisms, which cloud the water and reduce light penetration. This diminished visibility stresses aquatic life by hindering feeding and predator avoidance. Sublethal exposure to these chemicals can cause physiological and behavioral changes in aquatic organisms, affecting their reproduction and survival. For example, affected fish may have reduced reproductive capabilities, altered behaviors like nest abandonment, and greater vulnerability to disease and predation.

Certain pesticides can bioaccumulate in the tissues of aquatic organisms, posing risks to human health when contaminated fish are consumed. Some contaminants are known to cause serious health issues, including cancer and neurological disorders. Additionally, the uncontrolled use of pesticides can degrade food quality and pose various health risks to humans, highlighting the need for careful management and regulation of agricultural chemicals [68, 69].

4.2.2 Soil erosion and sedimentation

Soil erosion and the subsequent sedimentation in water bodies are significant environmental issues with far-reaching impacts on the capacity of these systems to store and filter water. This process begins when soil, primarily from agricultural lands subjected to intensive farming, deforestation, and inadequate land management practices, is eroded by wind and water. The topsoil layer, rich in essential nutrients for crops, is most vulnerable to this erosion. As the soil is carried away, soil fertility decreases on the land, negatively affecting crop yields and reducing the depth of arable soil. This challenges agricultural productivity and leads to the downstream transport of soil-laden water.

As eroded soil enters rivers, streams, and lakes, it causes sedimentation, accumulating soil particles that can significantly alter the aquatic environment. Sedimentation can obstruct water flow, reducing water velocity and potentially causing flooding. It also impacts the water bodies' ability to store water by decreasing their storage capacity, a crucial factor for water supply and hydropower generation. Furthermore, sedimentation impairs the natural filtration processes, affecting the water's physical, biological, and chemical qualities. Water quality degradation is exacerbated by the contaminants often bound to the soil particles, including heavy metals, nutrients, and pesticides from agricultural runoff. These contaminants threaten aquatic habitats, primary production, and human health by degrading drinking water quality.

The adverse effects of soil erosion and sedimentation extend beyond immediate water quality issues, influencing aquatic ecosystems' broader ecological balance and functionality. Suspended sediments can reduce sunlight penetration, affecting photosynthesis in aquatic plants and disrupting food webs. Additionally, the siltation of water reservoirs resulting from sediment accumulation can affect water supply reliability and the efficiency of hydropower generation by reducing water storage capacity [70–72].

4.2.3 Nutrient runoff and leaching

Eutrophication in water bodies, predominantly driven by nutrient runoff and leaching from agricultural lands, presents a critical environmental challenge. This process begins when excess nitrogen (N) and phosphorus (P) from animal manures and commercial fertilizers enter lakes, rivers, and streams. While essential for crop growth on land, these nutrients become pollutants when they enter aquatic environments in high concentrations. They act as fertilizers in these settings, too, but instead of supporting crops, they fuel the rapid growth of algae and aquatic plants.

Algal blooms, a direct consequence of eutrophication, can have several detrimental effects on water quality, aquatic life, and human health. These blooms might appear as nuisances, discoloring water and producing unpleasant odors. However, their impact goes much deeper. As algae grow unchecked, they block sunlight from reaching submerged plants, disrupting these plants' ability to photosynthesize and produce oxygen. This reduction in oxygen production is compounded when the algae eventually die and decompose. The decomposition process consumes significant dissolved oxygen in the water, leading to hypoxic conditions or oxygen depletion. This lack of oxygen creates a hostile environment for fish and other aquatic organisms, leading to die-offs and a reduction in biodiversity.

Moreover, some species of algae, mainly blue-green algae (cyanobacteria), can produce toxins harmful to aquatic and terrestrial animals, including humans. These toxins can cause a range of health issues, from skin rashes and gastrointestinal problems to more severe conditions such as liver damage and neurotoxic reactions. These toxins pose a risk to those who directly use or consume the contaminated water and animals and pets that may come into contact with or ingest the water.

Eutrophication also has broader ecological impacts. It can alter aquatic ecosystems to favor species tolerant of lower oxygen levels, thereby shifting the species composition and potentially introducing invasive species that can outcompete native flora and fauna. The increased plant growth and subsequent decomposition can further exacerbate the reduction in water quality, leading to a vicious cycle of degradation [73–76].

4.2.4 Economic implications of agricultural water pollution

The economic implications of agricultural water pollution and soil erosion are profound, spanning global to local scales and significantly impacting economies in various regions. Substantial resources are devoted to treating water contaminated by nutrients and pesticides to meet drinking standards. Eutrophication, mainly due to nutrient runoff into marine waters, imposes considerable economic burdens on commercial fisheries. Although a comprehensive assessment of the global economic costs is pending, individual studies shed light on the magnitude of these challenges.

In Denmark, efforts to reduce agricultural nutrient pollution have resulted in substantial financial expenditures and increased household water prices. The government's initiative to lower nitrogen leaching through improved practices and land use changes, such as afforestation and wetland creation, required farmers to cover 60% of the costs, totaling approximately $65 million. This initiative contributed to a 58% rise in water prices between 1988 and 1999, highlighting the economic impact of nutrient removal from water.

Similarly, the Netherlands faces significant challenges with agricultural pollution as a critical environmental issue, with the farming sector contributing heavily to water pollution levels. In the late 1990s, the annual external costs of eutrophication from nitrate emissions were estimated at around €600 million, with an additional €23 million annually spent on treating nitrates in drinking water.

In the UK, agriculture's role in water quality deterioration results in substantial economic costs, with farm runoff of nitrates, phosphorus, and pesticides being particularly problematic. The total annual cost of agricultural water pollution was estimated at around €725 million in the early 2000s, with nearly half of the water pollution prosecutions in the agricultural sector during 2002–2003 related to the dairy sector.

Globally, soil erosion contributes to these challenges, with estimated economic losses from reduced soil fertility, decreased crop yields, and increased water usage amounting to around $8 billion. In Java, Indonesia, soil erosion accounts for a 2% loss in total agricultural GDP, affecting farmers directly and others downstream. The U.S. agricultural sector faces about $44 billion per year in losses from erosion, including lost productivity, sedimentation, and water pollution, with lost farm income estimated at $100 million annually. Europe experiences $1.38 billion in annual agricultural productivity losses and $171 million in lost GDP due to soil erosion, while South Asia loses $10 billion annually.

These figures highlight the complex challenge in agricultural water pollution and soil erosion, with significant economic costs related to treating contaminated water, addressing eutrophication, and mitigating impacts on commercial fisheries and ecosystems. This underlines the urgent need for sustainable agricultural practices and NBSs to alleviate agricultural pollution's economic and environmental burdens [72, 77, 78].

4.3 Sustainable agriculture

Sustainable agriculture aims to meet present needs without compromising the ability of future generations to do the same. It involves careful stewardship of natural and human resources to ensure long-term productivity. This approach addresses resource depletion, environmental degradation, and social equity, making it essential to global sustainability initiatives.

By integrating site-specific plant and animal production practices, sustainable agriculture seeks to achieve long-term goals. These goals include meeting food and fiber demands, improving environmental quality, and ensuring economic sustainability for farms. Practices range from conventional to organic farming, designed to increase production for a growing population, conserve natural resources, and maintain economic viability.

Environmental protection is a priority in sustainable agriculture, which conserves natural resources and enhances soil fertility through responsible land management. It aims to sustain natural ecosystems and emphasizes social responsibility by ensuring safe and fair conditions for workers, addressing the needs of rural communities, and protecting consumer health and safety.

A holistic approach is central to sustainable agriculture. It recognizes the interconnectedness of farming, ecosystems, and communities. It promotes practices that support human and environmental well-being, striving to increase farm profitability, conserve the environment, enhance community quality of life, and boost production for consumption.

This approach reduces nonrenewable resources, favors renewable ones, and aligns with local sociocultural values. It aims to improve soil productivity, lessen environmental and health impacts, and promote sustainable employment and equitable conditions. Sustainable agriculture also fosters resilience against various challenges and supports the development of sustainable rural institutions through active stakeholder engagement.

Sustainable agriculture addresses the complex relationship between farming and water quality, blending ecosystem health with productivity. Tackling widespread agricultural pollution supports the sector's role in maintaining global ecosystem sustainability [79–83].

4.3.1 Key principles

The transition to sustainable food and agriculture is guided by five core principles: improving productivity, conserving the environment, fostering social equity, promoting economic growth, and meeting global food needs sustainably. These principles serve as a framework for protecting water quality and linking agriculture with environmental stewardship:

- *Increase productivity, employment, and value addition in food systems*: Enhancing agricultural productivity through water- and energy-efficient practices, such as precision agriculture (PA), can help minimize nutrient runoff and prevent water contamination, meeting rising crop demand while protecting aquatic ecosystems.
- *Protect and enhance natural resources*: Agroforestry and similar practices combine productivity with environmental protection. By integrating trees into farming, biodiversity increases, chemical runoff decreases, and water quality improves, helping to preserve natural habitats and protect water ecosystems.
- *Improve livelihoods and foster inclusive economic growth*: Sustainable agriculture aims to improve livelihoods and stimulate economic growth, particularly in rural areas. It emphasizes equitable access to resources and initiatives like microloans for women farmers, promoting sustainable farming and water conservation while addressing social equity and environmental care.
- *Enhance the resilience of people, communities, and ecosystems*: Resilience in sustainable agriculture is essential for coping with weather extremes, market changes, and social challenges. Climate-smart practices, such as drought-resistant crops and efficient irrigation, help adapt to climate shifts and conserve water, maintaining productivity and protecting water quality.
- *Adapt governance to new challenges*: Effective governance is crucial for sustainable agriculture, emphasizing accountability, equity, and transparency. Policies like watershed management promote practices that reduce chemical use, improve water quality, and integrate sustainability into agricultural regulations, supporting sustainable practices and protecting water resources [79–81].

4.3.2 Sustainable farming practices

Various sustainable farming practices are transforming agriculture by focusing on methods that enhance soil health, reduce dependency on chemical inputs, and improve ecosystem resilience.

4.3.2.1 Crop rotation and diversification

Crop rotation and diversification are vital practices in sustainable agriculture, crucial for preserving water quality and enhancing soil health. By varying the crops grown in succession, these methods optimize nutrient usage, reduce dependency on chemical inputs, and disrupt pest and pathogen cycles. Over time, diversifying crop species and traits improves soil structure and fertility, boosting resource use efficiency and ecosystem services. This approach supports agricultural biodiversity and leads to increased crop yields. Additionally, it contributes to better water management, as healthier soils have enhanced infiltration and reduced runoff, further safeguarding water quality and supporting sustainable farming objectives [84–86].

4.3.2.2 Organic farming

Organic farming is critical to sustainable agriculture, which enhances soil and ecosystem health by reducing synthetic inputs. By avoiding chemical fertilizers and pesticides, organic farming helps reduce the risk of groundwater pollution and supports aquatic health. It strengthens soil structure and improves water retention through practices such as crop rotation, the use of natural fertilizers, and biodiversity enhancement. This approach also reduces nutrient runoff into waterways and supports land and water ecosystems. Additionally, organic farming encourages biodiversity, decreases reliance on nonrenewable resources, and promotes ecological balance, contributing to the overall goals of sustainable agriculture and environmental stewardship [87–89].

4.3.2.3 Integrated pest management

Integrated pest management (IPM) is a critical strategy in sustainable agriculture, focused on reducing synthetic pesticides to protect the environment and water quality. IPM utilizes biological, cultural, and mechanical methods for pest control based on understanding pests' life cycles and their interactions with the environment. By establishing precise action thresholds, IPM enables timely and targeted interventions, minimizing the need for chemical use. This approach reduces the risk of pesticide runoff into waterways, helping to protect water resources and support biodiversity. IPM also contributes to maintaining ecological balance, essential for productive farming, and aligns with sustainability goals by promoting improved water quality, enhanced soil health, and reduced environmental impact [90–92].

4.3.2.4 Conservation tillage and cover crops

Conservation tillage, integrated with cover crop use, is pivotal in sustainable agriculture by enhancing water quality and soil health. Maintaining crop residues on the field significantly curtails erosion and augments water retention. Cover crops contribute further by enriching soil health, boosting organic matter, fostering biodiversity, improving crop yields, and providing greater weather resilience. These practices are economically viable and serve as a strategic investment, reducing operational costs and potentially increasing profits through better soil fertility. Conservation tillage and cover crops effectively reduce farming's ecological impact, supporting robust soil ecosystems and cleaner waterways and exemplifying essential sustainable farming practices focused on environmental stewardship and agricultural sustainability [93–98].

4.3.2.5 Agroforestry

Agroforestry integrates trees with crops and livestock, creating a sustainable agricultural model that boosts productivity and environmental health. Mimicking natural ecosystems improves soil health, reduces erosion, and enhances water retention. Practices like alley cropping, silvopasture, and riparian buffers increase biodiversity and reduce the need for artificial inputs while stabilizing ecosystems and providing economic benefits. Agroforestry strengthens agricultural resilience to climate change, making it a key strategy in sustainable farming [99–101].

4.3.2.6 Integrating livestock and crops

Integrating livestock and crops forms a cornerstone of sustainable agriculture, significantly enhancing water quality preservation. This holistic approach leverages livestock manure as a natural fertilizer, curtailing the need for chemical alternatives that risk water contamination. Conversely, crop residues serve as livestock feed, establishing a reciprocal relationship that bolsters soil health and structure. Such improved soil conditions increase water infiltration and decrease runoff, effectively reducing nutrient leaching into waterways. This creates a self-sustaining cycle that bolsters biodiversity and soil stability, which are critical factors in preventing erosion and protecting water resources. This integrated farming system promotes sustainable productivity and ensures water quality protection, demonstrating a comprehensive commitment to environmental stewardship within agriculture [102–104].

4.3.2.7 Precision agriculture

PA employs cutting-edge technology like GPS, automation, and sensors to boost agricultural efficiency and resource stewardship, supporting the objectives of sustainable farming and water conservation. It fine-tunes farming practices by closely monitoring and adjusting to variations within fields, enabling targeted management of crops. This precision minimizes excess use of water and fertilizers, thereby reducing chemical runoff into waterways and safeguarding water quality. Drones and in-ground sensors offer in-depth insights into soil, weather, and crop conditions, promoting prompt and accurate farming decisions. Additionally, precision livestock farming enhances this approach by tracking the health and productivity of animals, ensuring efficient use of resources across all farming activities [105, 106].

Case 4.1 Climate-Smart Agriculture Program, Australia

The Climate-Smart Agriculture Program, established by Australia's Natural Heritage Trust, signifies a pivotal $302.1 million commitment over 5 years starting in 2023–24. It aims to propel the agricultural sector toward sustainable, climate-resilient practices, enhancing productivity and competitiveness while embracing carbon and biodiversity incentives and industry sustainability frameworks. This initiative encourages farmers to integrate sustainable resource management practices, protecting and enriching natural capital and biodiversity. Central to the program are Partnerships and Innovation Grants, offering $45 million over 4 years for projects that foster climate-smart innovations and on-farm practices. Additionally, Capacity Building grants provide $15 million to enhance community capabilities in sustainable agriculture and natural resource management, emphasizing the success of existing projects and knowledge dissemination. Small Grants, with a $13 million fund, support grassroots projects by community and Landcare groups, promoting best practices in sustainable agriculture. Moreover, the Soil Capacity Building initiative focuses on fostering soil health knowledge and practices, with a substantial investment in Regional Soils Coordinators and a National Soils Community of Practice. This comprehensive program supports climate adaptation and mitigation efforts. It bolsters the agricultural sector's resilience, productivity, and environmental stewardship, aligning with global sustainability goals and enhancing the quality of life for farming communities [107].

Case 4.2 Ghana's Sustainable Agriculture Demonstration Farms

The Ministry of Food and Agriculture (MoFA) in Ghana has taken a significant step toward promoting sustainable agriculture and improving water quality by establishing 1,242 community demonstration centers across 117 districts. These centers serve as educational hubs, where farmers and stakeholders in the agricultural sector can learn about new, improved technologies in crop production, including drought-tolerant varieties of cassava, sweet potato, maize, rice, sorghum, groundnut, and cowpea. Initiated under the West Africa Agricultural Productivity Programme in 2013, these centers aim to increase the production of root and tuber crops, cereals, legumes, and livestock by generating and disseminating improved technologies. The project also encompasses community field demonstrations, training of trainers programs, study tours for farmers, and the production and distribution of technical materials to stakeholders. Approximately 171,229 farmers, including 68,492 women and 102,737 men, have benefited from this initiative. Training sessions have covered critical topics such as maize storage, yield loss assessment, crop production management, safe use of agrochemicals, access to credit, postharvest handling of vegetables, and food safety. These efforts aim to enhance agricultural productivity and contribute significantly to the preservation of water quality by promoting practices that reduce runoff and improve soil management [108].

4.4 Nature-based solutions for water quality preservation

NBS aim to protect, manage, and restore natural or modified ecosystems. They are designed to address societal challenges effectively and adaptively, benefiting human
well-being and biodiversity. In the context of agricultural landscapes, NBSs play a crucial role in enhancing water quality:
- *Water management*: NBSs can mimic natural processes and build on land restoration and operational water-land management concepts to improve vegetation,
  water availability, and quality and to raise agricultural productivity. They can enhance water quality and quantity for agricultural production while preserving
  the integrity of ecosystems.
- *Sustainable agriculture*: Agriculture NBS is a practical, long-term, cost-efficient approach to managing sustainable land and water resources and climate change. It
  can help improve water availability and quality and restore ecosystems and soils
  worldwide, offering substantial health co-benefits and achieving global food security.
- *Role of farmers*: Farmers, ranchers, and food producers are essential stewards of
  ecosystems and on the frontlines of climate change. They play an important role
  in developing and implementing environmental and agricultural solutions. They
  can use NBS to address water challenges while producing food.
- *Biodiversity and climate change*: When appropriately deployed, NBS can provide a
  triple benefit: improving the livelihoods of farmers and the resilience of agriculture, mitigating and adapting to climate change through soil, wetlands, and forests
  by carbon sequestration, and enhancing nature and biodiversity [36, 109, 110].

4.4.1 Integrating nature-based solutions into agricultural practices

NBSs offer a practical approach to addressing environmental challenges in agriculture, particularly those affecting water quality. Integrating NBS into agricultural practices provides a comprehensive strategy to reduce negative impacts on water bodies
while improving ecosystem health and agricultural productivity. This discussion explores the integration of NBS into agriculture, emphasizing sustainable practices,
green infrastructure, and conservation efforts.

4.4.1.1 Sustainable practices
NBS centers around sustainable agricultural practices that maintain and improve
water quality while supporting agricultural productivity. These practices include PA,
which uses technology to apply water and fertilizers more efficiently, thereby reducing runoff and leaching that may cause water pollution. Crop rotation and cover
crops enhance soil health, reducing erosion and the reliance on chemical inputs. By

avoiding synthetic fertilizers and pesticides, organic farming helps reduce groundwater and surface water contamination, contributing to healthier aquatic ecosystems.

4.4.1.2 Green infrastructure

Green infrastructure in agriculture refers to the strategic use of natural features and processes to provide environmental services, such as water purification and soil stabilization. Constructed wetlands, for example, can filter pollutants from agricultural runoff before it reaches water bodies. Buffer vegetation zones along waterways trap sediment and absorb nutrients and pesticides, reducing the load entering streams and rivers. These green infrastructure elements work with natural landscapes to mitigate pollution, stabilize soil, and improve water infiltration.

4.4.1.3 Amelioration

Amelioration focuses on restoring degraded agricultural lands and improving their capacity to support healthy ecosystems. Techniques such as restoring riparian zones and implementing soil health practices can reverse the detrimental impacts of intensive agriculture. By enhancing the natural filtration capacity of soils and vegetation, these practices contribute to cleaner waterways. Restoring wetlands within agricultural landscapes improves water quality, enhances biodiversity, and provides habitat for wildlife.

4.4.1.4 Conservation

Agriculture conservation efforts aim to protect and enhance biodiversity, ecosystem connectivity, and natural carbon sinks. Conservation practices bolster ecosystem resilience to climate change and reduce reliance on chemical inputs by maintaining or restoring natural habitats within and around agricultural lands. Agroforestry, the integration of trees and shrubs into farming systems, exemplifies conservation. It creates diverse agricultural landscapes that mimic natural ecosystems, offering benefits such as enhanced water retention, reduced erosion, and improved habitat for beneficial organisms. These efforts collectively contribute to carbon sequestration in soils and vegetation, mitigating climate change impacts and improving water quality [36, 110, 111].

4.4.2 Specific NBS approaches in agriculture

Various NBS approaches can be implemented in agricultural settings to protect water quality, including the following.

4.4.2.1 Constructed wetlands

Constructed wetlands are engineered ecosystems designed to filter and treat agricultural runoff, improving water quality before reentering the natural water cycle. These systems intercept runoff from agricultural lands, allowing water to pass through vegetation, soil, and microbial communities that work together to remove pollutants like nutrients, sediments, pesticides, and pathogens.

Plants in constructed wetlands absorb excess nutrients and heavy metals, while root systems and microbes break down organic matter and contaminants. The soil acts as a filter, trapping sediments and further reducing pollutant levels. This process cleans runoff water and supports biodiversity by creating habitats for aquatic and terrestrial species.

Constructed wetlands provide an effective method for managing agricultural runoff, reducing the pollutants entering rivers, lakes, and groundwater, and supporting the protection of aquatic ecosystems. These systems also contribute to carbon sequestration, flood mitigation, and enhanced landscape aesthetics, positioning them as a valuable element of NBS in agriculture [36].

4.4.2.2 Riparian buffers

Riparian buffers are vegetated areas along the margins of water bodies, such as rivers, streams, and lakes, designed to intercept pollutants from agricultural runoff before reaching these aquatic ecosystems. These buffers consist of grasses, shrubs, and trees that act as a natural barrier, filtering out sediments, nutrients, pesticides, and other contaminants from runoff water. The vegetation in riparian buffers plays a critical role in absorbing excess nutrients and chemicals, while their roots stabilize soil, reducing erosion and sediment transport into waterways.

The strategic placement of riparian buffers enhances water quality by slowing water flow, allowing pollutants to settle out and be absorbed or broken down by plants and soil microbes. This process effectively reduces the amount of pollutants entering water bodies, protecting aquatic habitats and improving the health of aquatic organisms. Riparian buffers also contribute to biodiversity by providing habitat and travel corridors for various wildlife species, including birds, mammals, and insects.

Implementing riparian buffers is a cost-effective and environmentally friendly approach to mitigate the adverse effects of agricultural practices on water quality. These natural filtration systems are an essential component of integrated water resource management, offering a sustainable solution for preserving the integrity of water bodies while supporting agricultural productivity and ecosystem services [36].

4.4.2.3 Agroforestry systems

Agroforestry systems, which integrate trees with crops and/or livestock, offer a sustainable approach that merges agricultural productivity with ecosystem conservation, enhancing water infiltration and quality. These systems mimic natural ecosystems, utilizing the beneficial interactions between plant species and the soil to create a more resilient agricultural landscape. Trees and shrubs in agroforestry systems are pivotal in improving soil structure, increasing organic matter, and enhancing the soil's ability to absorb and retain water. This results in improved water infiltration rates, reducing surface runoff and the potential for erosion, thereby lessening the transport of pollutants into nearby water bodies.

By increasing the permeability of the soil, agroforestry systems help recharge groundwater supplies and maintain stream flows, contributing to better water availability for both agricultural and community use. Trees' deep root systems are particularly effective in accessing water from deeper soil layers, which can be beneficial in drought conditions and nutrient cycling, reducing the need for synthetic fertilizers.

Furthermore, integrating trees and vegetation buffers within agricultural landscapes can filter out pollutants before they reach water bodies, significantly improving water quality. Agroforestry practices support biodiversity, offering habitats for various species, and play a crucial role in carbon sequestration, helping to mitigate climate change impacts. This holistic approach sustains agricultural productivity and promotes the conservation of vital water resources and ecosystem services [36].

4.4.2.4 Biofiltration systems

Biofiltration systems are innovative solutions designed to manage nutrient runoff from agricultural lands, significantly reducing the reliance on synthetic fertilizers. These systems employ natural or engineered living materials, such as plant roots, soil microbes, and organic matter, to filter out pollutants and excess nutrients from runoff water before they can enter water bodies. By capturing and breaking down nitrates, phosphates, and other harmful nutrients, biofiltration systems prevent them from causing eutrophication in rivers, lakes, and estuaries, which can lead to detrimental algal blooms and hypoxic zones detrimental to aquatic life.

The process involves strategically placing biofiltration media, such as constructed wetlands, vegetated swales, or riparian buffer zones, in critical areas where agricultural runoff is most likely to occur. These systems trap sediments and nutrients and provide a habitat for a diverse range of microorganisms that naturally break down pollutants, further enhancing water quality.

Biofiltration systems contribute to sustainable agriculture by improving soil health and reducing the need for chemical inputs. Enriched with organic matter and beneficial microbes, soils become more fertile and productive, thus lowering the demand for synthetic fertilizers. This eco-friendly approach to nutrient management supports the circular economy within agriculture, turning potential pollutants into valuable resources for enhancing soil fertility. By integrating biofiltration systems into agricultural practices, farmers can protect water quality while maintaining or even increasing crop yields, demonstrating a commitment to environmental stewardship and sustainable farming [36].

Case 4.3 Toolkit for estimating constructed wetland size for farm runoff treatment

The Wildfowl & Wetlands Trust has developed the Farm Wetland Sizing Toolkit for Catchment Sensitive Farming, providing farmers with a resource to estimate the size needed for constructed wetlands aimed at treating lightly contaminated yard runoff. This toolkit, designed to support the Countryside Stewardship constructed wetland option, is instrumental in the initial planning stages for farms exploring natural solutions to manage runoff. It focuses on a three-stage wetland system, specifically for pollution treatment scenarios not involving high-strength pollutants such as silage effluent, slurry, dairy washings, or septic tank outflow. While the toolkit offers a preliminary estimation, it emphasizes the need for a comprehensive farm water management plan for detailed design, accounting for the specific composition and volume of inflow wastewater. This approach aids farmers in integrating sustainable practices into their operations, contributing to improved water quality and aligning with broader environmental conservation goals [112].

Case 4.4 AGROMIX: pioneering agroforestry and mixed farming systems for European resilience and efficiency

The AGROforestry and MIXed farming systems project, supported by Horizon 2020, aims to catalyze a transition toward more resilient and efficient land use across Europe. Focused on implementing agroecological practices within farm and land management, this project targets integrating mixed farming and agroforestry systems. With activities from 2020 to 2024, AGROMIX engages in participatory research to uncover and harness synergies between these systems, promoting sustainable value chains and developing practical toolkits for codesigning and managing agroecological solutions. This initiative is grounded in a multiactor approach. Twelve pilot sites collaborate on codesigning systems and pathways for transition. Additionally, nine experimental sites are dedicated to measuring resilience indicators, modeling transition scenarios, and refining methods for greenhouse gas inventories. Among its innovative strategies, AGROMIX is developing a serious game to facilitate knowledge sharing and learning among stakeholders. The project's comprehensive activities include economic analyses, value chain development, policy recommendations, and action plans to foster impactful transitions. With a broad network of 83 case study sites and extensive dissemination efforts, AGROMIX seeks to impact farms within value chains, across policy levels significantly, and throughout Europe's agricultural sector, contributing to the goals of low-carbon, climate-resilient farming [113].

Chapter 5
Integrated water and nutrient reuse and recycling in sustainable agriculture

Abstract: This chapter explores the vital roles of water reuse and recycling within the framework of sustainable agriculture, which is essential for conserving finite water resources amid growing demands and climate change challenges. It delves into the principles of circular food systems, integrating sustainable water management strategies to minimize waste, enhance resource efficiency, and support ecosystem resilience. The discussion includes various sources of water for reuse and recycling, advanced treatment technologies, and their applications in agriculture, such as irrigation, aquaponics, and greenhouse farming. The chapter also introduces integrated water and nutrient recycling systems, highlighting technologies that enable the simultaneous recovery and reuse of water and essential nutrients.

5.1 Introduction

Water reuse and recycling are essential for sustainable agriculture within a circular food system. Water reuse involves repurposing treated wastewater for irrigation, aquaculture, industrial processes, or groundwater recharge. Water recycling in agriculture includes collecting, treating, and reusing water on the farm, such as capturing rainwater or reusing water in hydroponic setups, to reduce reliance on external water sources. These practices aim to decrease demand for freshwater, lower pollution levels, and reduce greenhouse gas emissions, thereby supporting food security, promoting biodiversity, and enhancing the resilience of agricultural systems against climate change. Integrated water and nutrient recycling systems further these efforts by recovering and reusing water and essential nutrients from agricultural and wastewater streams. These systems use advanced technologies to improve resource efficiency, reduce waste, and support sustainable agricultural practices by providing a consistent supply of recycled water and nutrients.

This chapter discusses the crucial roles of water reuse and recycling in modern agriculture, which accounts for approximately 70% of the world's freshwater usage. It analyzes the current landscape, including the benefits, challenges, and emerging opportunities, to examine how these water management strategies are vital for advancing sustainable agricultural practices globally [114].

https://doi.org/10.1515/9783111341385-005

5.2 Introduction to circular food systems and water management

Integrating circular food systems with sustainable water management significantly minimizes waste and optimizes resource use throughout the food production and consumption cycle. This section explores the principles and practices of circular food systems, defining their core concepts and highlighting the essential role of efficient water use in achieving circularity. Strategies for recycling and reusing water are examined to demonstrate how resilient, sustainable, and equitable food systems can be developed to benefit businesses, society, and the environment.

5.2.1 Defining circular food systems

Circular food systems are predicated on the principles of the circular economy, which aims to redefine growth by focusing on positive society-wide benefits. This approach employs practices that seek to eliminate waste and continually use resources within food systems, creating a regenerative system.

In circular food systems, all stages of production and consumption are optimized for efficiency and sustainability, from sourcing materials to managing by-products and waste. Unlike traditional linear models, circular models employ recycling, composting, and anaerobic digestion strategies to ensure that organic waste is returned to the system as a nutrient source for crops or used to generate energy.

A core component of the circular economy in food systems is the innovative use of organic resources. Food by-products are viewed not as waste but as valuable resources that can be reintegrated into the cycle. These can be transformed into organic fertilizers, new food products, materials for the fashion industry, or bioenergy sources, reducing the pressure on landfills and the need for synthetic fertilizers, thereby promoting soil health and biodiversity.

Sustainable water management is integral to achieving circularity in food systems. Water, a finite and vital resource, is heavily utilized in agriculture and food production. Circular food systems advocate for efficient water use through rainwater harvesting, wastewater treatment and reuse, and precision irrigation technologies. These systems minimize waste and lessen the impact on freshwater resources by ensuring water is used and reused efficiently. This is crucial for the sustainability of food systems and the resilience of ecosystems and communities dependent on these water sources.

Circular food systems represent a holistic and sustainable food production and consumption management approach. They aim to close the loop on waste, create efficiencies in resource use, and maintain ecosystems' health. By embedding circular economy concepts into food systems and prioritizing sustainable water management, pathways toward more sustainable, resilient, and equitable food futures can be forged [13, 115, 116].

5.2.2 Roles of water reuse and recycling

Water reuse and recycling are vital strategies in sustainable water management within circular food systems. Though often used interchangeably, these practices serve distinct roles in conserving water resources, each contributing uniquely to reducing water wastage and enhancing overall sustainability.

Water reuse, or water reclamation, involves capturing, treating, and using wastewater from various sources for beneficial purposes. These include agricultural irrigation, industrial processes, groundwater recharge, and potable uses following adequate treatment. The primary goal of water reuse is to extend the utility of water by enabling its repeated use for different applications, thereby reducing the demand for freshwater sources. This practice is particularly crucial in regions facing water scarcity, as it provides an alternative water source, lessening the strain on overexploited natural water bodies.

Meanwhile, water recycling refers to treating and using wastewater within the same system or facility where it was produced. For example, in a manufacturing plant, water recycling might involve collecting, treating, and reintegrating the processed water back into the production cycle. This closed-loop system minimizes water withdrawal from external sources and significantly reduces the water footprint. Recycling can take the form of capturing and reusing runoff or implementing soil moisture conservation techniques to optimize water use in agriculture.

Both practices are integral to minimizing water wastage. Treating and reusing water significantly reduce the volume of wastewater discharged into the environment, mitigating pollution and conserving water. This is vital not only for protecting aquatic ecosystems and biodiversity but also for ensuring the availability of clean water for future generations.

The importance of water reuse and recycling extends beyond conservation. These practices are essential to a broader strategy to achieve water sustainability, particularly in growing global challenges such as climate change, population growth, and urbanization. Integrating water reuse and recycling into circular food systems make more resilient and efficient water management frameworks possible. These frameworks support sustainable food production and contribute to the broader goals of economic development, social well-being, and environmental stewardship [13–15].

5.3 Water reuse in agriculture: sources, technologies, and applications

With increasing pressures on global water resources due to climate change and growing agricultural demands, water reuse in agriculture plays an essential role in managing water scarcity and maintaining productivity. Water reuse helps conserve freshwater supplies while supporting agricultural practices. This section examines the various sources of water that can be reused in agricultural settings, the technologies that facilitate effective treatment and recycling, and the practical applications of these methods in agriculture.

5.3.1 Sources of water for reuse

In sustainable water management, identifying and utilizing various water sources for reuse are critical. These sources encompass municipal water, industrial water, agricultural water, and stormwater, each presenting unique qualities and treatment needs to meet reuse standards.

5.3.1.1 Municipal wastewater

Municipal wastewater originates from household, commercial, and institutional use. It includes water from sinks, showers, and toilets, which contain a mix of biological and chemical pollutants. Before reuse, municipal wastewater requires comprehensive treatment to remove contaminants and pathogens to ensure it meets safety standards for its intended use, such as agricultural irrigation or industrial processes.

5.3.1.2 Industrial wastewater

Industrial wastewater is the by-product of manufacturing and industrial activities. Its composition varies widely depending on the industry, ranging from relatively clean water to highly polluted effluents containing chemicals, heavy metals, and organic pollutants. Treating industrial wastewater for reuse necessitates tailored solutions that address the contaminants present, ensuring the water is fit for reuse within the same facility or in other applications.

5.3.1.3 Agricultural runoff

Agricultural runoff, another significant source, includes water used for irrigation and carries fertilizers, pesticides, and organic matter. Collecting and treating this runoff can provide a valuable water source for reuse in irrigation, reducing the reliance on fresh water and mitigating nutrient pollution in water bodies.

5.3.1.4 Stormwater

Stormwater, the runoff from rain or snowmelt, presents an underutilized resource. While relatively clean compared to other sources, stormwater can collect pollutants from urban surfaces. Capturing and treating stormwater can mitigate urban flooding risks and supply additional water for non-potable uses, such as watering green spaces or industrial cooling.

Each of these water sources requires careful assessment of water quality and the development of appropriate treatment processes to meet the specific requirements of their intended reuse applications. This ensures the safety and sustainability of water reuse practices, contributing to the overall efficiency of water resource management [13–15].

5.3.2 Technologies for water reuse

Technological advancements have significantly enhanced the efficiency and safety of water reuse processes. The core treatment methods include filtration, biological treatment, and disinfection, each vital in preparing wastewater for safe reuse across various applications.

5.3.2.1 Filtration

Filtration processes are fundamental in removing suspended solids, particles, and certain types of pollutants from water. Techniques range from simple sand filtration, which captures physical impurities, to more sophisticated membrane filtration, such as ultrafiltration and reverse osmosis. Membrane technologies are particularly effective in removing many contaminants, including bacteria, viruses, and dissolved salts, making them crucial for producing high-quality reusable water.

5.3.2.2 Biological treatment

Biological treatment leverages microorganisms to degrade organic matter and pollutants present in wastewater. This method is typically employed in the initial stages of wastewater treatment and involves processes like activated sludge, biofilms, and aerated lagoons. These biological systems are designed to mimic natural purification processes, efficiently breaking down organic pollutants into harmless by-products.

5.3.2.3 Disinfection

Disinfection is the final step, crucial for ensuring the water is free from pathogenic microorganisms. Chlorination is a widely used disinfection method, effective against a broad spectrum of pathogens. However, concerns over chemical residues have led to adopting alternative methods, such as ultraviolet irradiation and ozonation, which offer effective disinfection without adding chemicals.

5.3.2.4 Innovations in water reuse technologies

Innovations in water reuse technologies focus on improving efficiency and safety while minimizing environmental impact. Advances in membrane technology, for instance, are making filtration processes more energy-efficient and effective at removing a more comprehensive range of contaminants. Additionally, developments in biological treatment methods are enhancing the degradation of pollutants, even in challenging wastewater compositions. These technological advancements are crucial for expanding the potential of water reuse and offering sustainable solutions for managing water resources in an increasingly water-scarce world [117–120].

5.3.3 Applications in agriculture

Reusing treated wastewater in agriculture is increasingly recognized as a sustainable solution to meet the water needs of crops, aquaculture, and other agricultural practices. This approach conserves valuable freshwater resources and contributes to nutrient recycling, enhancing soil fertility and productivity.

5.3.3.1 Irrigation
In irrigation, reused water provides a reliable water source for crop cultivation, especially in arid and semiarid regions where water scarcity is a significant challenge. For example, treated municipal wastewater is commonly used for irrigating nonedible crops such as cotton and ornamental plants, and edible crops in some systems where regulations and treatment levels ensure safety. The nutrients present in the treated water, particularly nitrogen and phosphorus, can reduce the need for synthetic fertilizers, lowering production costs and minimizing environmental pollution.

5.3.3.2 Aquaculture: farming fish and aquatic plants
Aquaculture, the farming of fish and aquatic plants, also benefits from reused water. In closed-loop aquaculture systems, water is treated and recycled, reducing the need for continuous freshwater inputs and minimizing the discharge of pollutants into natural water bodies. This sustainable approach allows for the expansion of aquaculture operations even in areas with limited access to freshwater.

5.3.3.3 Other agriculture applications
Other agricultural applications of reused water include livestock watering and the processing of agricultural products. Reused water can be treated to appropriate quality standards to ensure it is safe for livestock, thereby alleviating the pressure on freshwater sources. Additionally, it can be used in various stages of food processing, such as washing and cooling, provided that it meets the necessary safety standards [13–15].

Case 5.1 SA Water – Northern Adelaide Irrigation Scheme (NAIS)

SA Water's NAIS is a landmark initiative in water recycling and sustainable agriculture. As Australia's second largest water recycler, SA Water reuses one in every 3 L of treated wastewater, supporting various agricultural and horticultural activities. The NAIS, funded by the Australian Government and the Government of South Australia, is designed to unlock 12 GL of high-quality recycled water annually. This water supports over 300 ha of high-technology horticulture and an additional 2,700 ha of advanced agri-food production. The scheme is vital for irrigating horticulture, floriculture, fruit and nut orchards, table and wine grapes, and high-value broad-acre cropping. The scheme also supports poultry and other intensive animal husbandry operations. Critical infrastructure investments totaling $155.6 million include a wastewater treatment plant that produces 6 GL/year of high-quality recycled water, extensive seasonal balancing storage, and a transmission main from Bolivar to the Northern Adelaide Plains. The distribution network includes spur lines and connection points to the farm gate, with pumping stations as needed. This infrastructure ensures a long-term supply of reliable, climate-independent water at stable prices, fostering efficient and sustainable agricultural practices. NAIS has generated significant economic benefits, creating 6,000 jobs, attracting $2 billion in private investment, and contributing over $1 billion annually to the state's economy. The scheme's pricing structure includes a one-off capital contribution for connection, an annual availability charge based on contracted water volumes, and a consumption charge based on actual water use. These charges are indexed annually to stabilize prices and protect participants from unexpected cost increases. Water rights within the scheme are fully tradable, allowing long-term water contracts to be sold or transferred, subject to system capacity and regulatory approval. This flexibility supports agribusiness growth and ensures efficient water use across the Northern Adelaide Plains, making it a hub for developing export agri-food businesses. By leveraging advanced water recycling technologies and strategic investments, SA Water's NAIS exemplifies sustainable water management, supporting the region's agricultural productivity and economic growth [121, 122].

5.4 Water recycling in agriculture: sources, technologies, and applications

Water recycling in agriculture is a crucial practice for enhancing the sustainability of water use, particularly in regions facing water scarcity. By recycling water within agricultural systems, it is possible to reduce reliance on external water sources and improve overall water efficiency. This section explores the various sources of water that can be recycled in agricultural settings, the technologies that enable effective recycling, and the practical applications of these methods in agriculture. Water recycling practices can help optimize water use in agriculture, contributing to more sustainable and resilient farming systems.

5.4.1 Sources of water for recycling in agriculture

Agriculture can benefit from various unconventional water sources for recycling, such as rainwater, fog, dew, and runoff. Utilizing these sources helps reduce dependence on traditional water supplies and enhances sustainability in agricultural practices.

5.4.1.1 Rainwater
Rainwater is a valuable resource for agriculture, capable of being captured from various surfaces and stored for later use. It is relatively clean and, with minimal treatment, can be used for irrigation and watering livestock. Techniques for capturing and storing rainwater include the following:
– Installing collection systems on greenhouse roofs and other farm structures to funnel water into storage tanks
– Creating rainwater harvesting ponds and reservoirs that collect runoff for large-scale irrigation purposes

5.4.1.2 Fog
Fog collection utilizes the moisture from fog, converting it into usable water through condensation. This method is particularly beneficial in arid regions where fog is more common than rain. Techniques for capturing fog for agricultural use involve the following:
– Setting up fog nets or mesh systems on farms to capture moisture, which condenses and drips into collection troughs or tanks
– Strategically placing these systems in areas with high fog incidence to maximize water collection

5.4.1.3 Dew
Dew, while offering smaller quantities of water, can still contribute to meeting crops' water needs. Dew collection systems capture moisture from the air, providing supplementary water for irrigation. Techniques include the following:
– Utilizing condensation traps or sheets placed close to the ground to collect dew during the cooler parts of the night
– Implementing innovative materials designed to enhance dew collection efficiency and direct water to the root zones of plants [13–15]

5.4.2 Technologies for water recycling in agriculture

Advanced technologies for water recycling in agriculture play a crucial role in optimizing water use and promoting sustainability. These technologies enable efficient water management by reducing waste and enhancing water reuse within agricultural systems.

5.4.2.1 Drip irrigation

One of the most water-efficient methods, drip irrigation, delivers water directly to the root zone of plants, minimizing evaporation and runoff. This technology allows for the precise application of water, ensuring that crops receive the exact amount needed for optimal growth. Innovations in drip irrigation include:

- The integration of water-soluble fertilizers into the irrigation system enables fertigation, which optimizes nutrient uptake and reduces water usage.
- Smart controllers adjust watering schedules based on real-time soil moisture data, weather forecasts, and plant water requirements.

5.4.2.2 Soil moisture conservation methods

Techniques such as mulching, cover cropping, and superabsorbent polymers reduce soil moisture evaporation and improve water retention. These practices contribute to maintaining soil moisture levels, reducing the need for frequent irrigation. Key advancements include:

- The development of biodegradable mulches that enrich soil health while conserving moisture
- Using precision agriculture tools to map soil moisture variability that allows for targeted conservation efforts

5.4.2.3 Closed-loop systems

These systems recycle all water used in agricultural processes, minimizing waste and maximizing efficiency. In closed-loop aquaponics and hydroponics systems, water is continuously cycled between fish tanks and plant growing areas, with plants filtering and cleaning the water for reuse. Recent advancements in closed-loop systems include:

- Automating water monitoring and recycling processes ensures optimal water quality and reduces labor costs.
- Incorporating renewable energy sources to power water pumps and filtration units further enhances the sustainability of these systems [13–15].

5.4.3 Applications in agriculture

Water recycling has practical applications across various agricultural practices, helping to improve water efficiency and support sustainable farming. This section explores how recycled water is used in agriculture, highlighting its role in irrigation, aquaponics, hydroponics, and other vital areas.

5.4.3.1 Irrigation efficiency

Using recycled water in irrigation is essential for improving water efficiency in agriculture. By utilizing treated wastewater and other recycled water sources, farmers can reduce their dependence on freshwater supplies, which is particularly important in regions facing water scarcity. Recycled water can provide a reliable and sustainable source for crop irrigation, ensuring consistent water availability throughout growing seasons. Additionally, the nutrients present in recycled water can supplement soil fertility, potentially reducing the need for chemical fertilizers.

5.4.3.2 Aquaponics and hydroponics

Aquaponics and hydroponics are innovative agricultural practices that benefit significantly from water recycling. In aquaponics, a symbiotic system is created where fish waste provides essential nutrients for plant growth, and the plants, in turn, help filter and clean the water, which is then recirculated back into the fish tanks. Hydroponics involves growing plants in a nutrient-rich water solution without soil. Both systems are highly efficient in water use, requiring significantly less water than traditional soil-based agriculture. Recycling water in these systems conserves water and allows for the continuous reuse of nutrients, making these methods particularly suitable for controlled environment agriculture (CEA).

5.4.3.3 Greenhouse farming

Modern greenhouse farming increasingly incorporates advanced water recycling systems to enhance water efficiency and sustainability. These systems are designed to capture and reuse condensation and runoff, significantly reducing water waste and ensuring a consistent water supply for plants. By recycling water within the greenhouse, farmers can optimize water use, particularly in controlled environments where precision is crucial. Additionally, many greenhouses integrate water treatment technologies to maintain high water quality, which is essential for promoting healthy plant growth and maximizing crop yields.

5.4.3.4 Rainwater harvesting for agriculture

Rainwater harvesting is a practical and sustainable method used in agriculture to supplement water supplies, particularly in areas with limited access to freshwater resources. This technique involves collecting and storing rainwater from rooftops, fields, or other surfaces, which can be used for irrigation during dry periods. Farmers can reduce reliance on external water sources by capturing rainwater, ensuring a more consistent and self-sufficient water supply. Rainwater harvesting systems can range from simple collection tanks to more complex setups with filtration and distribution networks, allowing for efficient use of the collected water [4, 13–15].

Case 5.2 Drip irrigation systems fed by rainwater harvesting in Jamaica
Jamaica has implemented advanced drip irrigation systems fed by rainwater harvesting to enhance water efficiency in agriculture. These systems aim to address the challenges posed by climate change, such as prolonged droughts and irregular rainfall patterns, which have significantly impacted the agricultural sector. Drip irrigation technology involves the precise application of water and minerals directly to the root zones of crops, minimizing water loss through deep percolation or evaporation. This method significantly increases water use efficiency and crop yields while reducing wastage. The system benefits rows, fields, and tree crops grown closely together. Integrating rainwater harvesting with drip irrigation also provides a reliable water supply, reducing dependence on inconsistent rainfall. One initiative distributed 2,000 drip irrigation systems to small farmers across Jamaica in 2020. This project aimed to ensure the continued growth of the agricultural sector despite changing rainfall patterns due to climate change. The Yallahs-Hope On-Farm Water Management Project further equipped 315 farmers with rainwater harvesting sheds and drip irrigation systems, enhancing their irrigation capabilities. The benefits of these technologies include providing water to farms as needed, conserving groundwater, and reducing the risk of saline intrusion in coastal aquifers. Drip irrigation also improves the efficiency of chemical fertilizer application through fertigation, preventing resource waste and minimizing environmental impacts such as waterway pollution and biodiversity loss. The systems reduce soil degradation and erosion associated with traditional irrigation methods, preserving water sources and reducing siltation. Despite the advantages, the uptake by small and medium-sized farmers using these technologies remains low. Efforts to increase adoption include providing training on installation, operation, and maintenance and data collection to monitor water use and crop yields. These initiatives aim to promote the benefits of drip irrigation and rainwater harvesting, encouraging more farmers to implement these sustainable practices [123].

5.5 Integrated water and nutrient recycling systems

Integrated water and nutrient recycling systems represent a significant advancement in sustainable agriculture, enabling the simultaneous recovery and reuse of water and essential nutrients. These systems are designed to enhance resource efficiency, reduce waste, and support sustainable farming practices by treating and recycling agricultural and wastewater streams. This section explores the technologies that drive these systems, their applications in agriculture, and the economic and environmental benefits they offer.

5.5.1 Technologies for simultaneous water and nutrient recovery

Simultaneous water and nutrient recovery technologies are essential for advancing sustainable agricultural practices. These innovations allow for the efficient treatment and recycling of water and nutrients from agricultural and wastewater sources, enhancing resource use and minimizing waste. This section provides an overview of critical technologies that enable integrated water and nutrient recycling, supporting the development of more resilient and sustainable farming systems.

5.5.1.1 Membrane bioreactors
Membrane bioreactors combine biological treatment processes with membrane filtration to treat wastewater and recover nutrients. The biological treatment decomposes organic matter while the membranes filter out suspended solids, pathogens, and other contaminants. This dual-action system produces high-quality effluent suitable for reuse in irrigation and recovers valuable nutrients like nitrogen and phosphorus, which can be reused as fertilizers.

5.5.1.2 Anaerobic digestion
Anaerobic digestion involves the microbial decomposition of organic matter without oxygen, producing biogas (a renewable energy source) and digestate rich in nutrients. The digestate can be further processed to extract nutrients for agricultural use, while the treated water is suitable for irrigation or other non-potable uses. This process not only recovers nutrients but also contributes to energy production.

5.5.1.3 Constructed wetlands
Constructed wetlands are engineered systems that mimic natural wetlands to treat wastewater. These systems use vegetation, soil, and microbial activity to remove contaminants and recover nutrients. Plants uptake nutrients as wastewater flows through the wetland, and microorganisms break down organic matter. The treated water can be reused, and plant biomass can be harvested as organic fertilizer.

5.5.1.4 Nutrient film technique
The nutrient film technique is a hydroponic system where a thin film of nutrient-rich water flows over the roots of plants. This method allows precise control over nutrient

delivery, ensuring optimal plant growth while minimizing water and nutrient waste. The excess nutrient solution can be recirculated, making the system highly efficient in water and nutrient use.

5.5.1.5 Struvite precipitation

Struvite precipitation is a chemical process that recovers phosphorus and ammonia from wastewater by forming struvite crystals, which can be used as a slow-release fertilizer. This method effectively reduces nutrient loads in wastewater and provides a valuable fertilizer product for agricultural use.

Case 5.3 Waternet – Airprex at the wastewater treatment plant (WWTP) Amsterdam West

Waternet, the water utility for Amsterdam, implemented the Airprex process at WWTP Amsterdam West to enhance sewage sludge treatment and recover phosphorus. The WWTP handles wastewater from 1 million population equivalents and sludge from 2 million, producing 13 million m^3 of biogas annually. The primary challenge at WWTP Amsterdam West was the significant buildup of struvite crystals, leading to scaling in pipelines and dewatering equipment. This problem was addressed by the Airprex process, which facilitates struvite crystallization through pH rise via CO_2 stripping and the addition of magnesium chloride ($MgCl_2$). The pilot phase of Airprex demonstrated significant improvements. Phosphate levels dropped from 150 to 5 mg/L after crystallization, and sludge dewaterability increased from 22% dry matter (DM) to 25–26% DM. In addition, the polymer dosage required for dewatering was reduced. Economically, the Airprex process yielded substantial savings. The annual benefits amounted to approximately €1.2 million, with costs around €700,000, resulting in net annual savings of €500,000. The total investment cost was €3 million, with a 6-year return on investment. The process improvements included the elimination of scaling issues and enhanced struvite production. In 2018, struvite production increased from 200 tons in 2017 to 300 tons, with a projection of 500 tons by 2025. The recovered struvite, tested by ICL Fertilizer, was of high quality and suitable for use in tailored fertilizers. Overall, Waternet's Airprex process at WWTP Amsterdam West successfully addressed the phosphorus problem, improved operational efficiency, and provided a valuable fertilizer product, showcasing a model for sustainable wastewater management [124].

5.5.2 Applications in sustainable agriculture

Integrating water and nutrient recycling technologies is pivotal in advancing sustainable agriculture. These applications enhance resource efficiency, improve crop productivity, and reduce environmental impact. This section highlights how these integrated systems are applied in various agricultural practices, contributing to the overall sustainability and resilience of farming operations.

5.5.2.1 Irrigation

Incorporating recycled water from integrated water and nutrient recycling systems into irrigation practices offers significant benefits for sustainable agriculture. Farmers can reduce their dependence on freshwater resources by using nutrient-rich recycled water for irrigation while supplying essential nutrients to crops. This dual-purpose approach enhances water use efficiency and promotes soil fertility, potentially lowering the need for synthetic fertilizers. The consistent availability of recycled water also helps stabilize crop production, particularly in regions facing water scarcity.

5.5.2.2 Hydroponics and aquaponics

Hydroponic and aquaponic systems benefit significantly from integrating water and nutrient recycling technologies. In hydroponics, plants are grown in a nutrient-rich water solution without soil, allowing for precise control over nutrient delivery and water use. This system recycles water continuously, minimizing waste and reducing the need for freshwater inputs. Aquaponics combines hydroponics with aquaculture, creating a closed-loop system where fish waste provides nutrients for plant growth, and the plants help filter and clean the water before it is recirculated back to the fish tanks. Integrating water and nutrient recycling in these systems conserves water and enhances nutrient use efficiency, leading to higher productivity and sustainability. These methods suit controlled environments well, enabling year-round production and contributing to more resilient and resource-efficient agricultural practices.

5.5.2.3 Fertigation

Fertigation, the practice of delivering fertilizers through irrigation systems, is further enhanced by using nutrient-rich recycled water. By combining irrigation with fertilization, fertigation ensures that crops receive water and essential nutrients simultaneously, optimizing nutrient uptake efficiency. This integrated approach reduces the need for separate fertilizer applications, thereby minimizing the risk of fertilizer runoff into surrounding environments. Using recycled water in fertigation also supports more sustainable nutrient management practices by making the most available resources and reducing reliance on synthetic fertilizers.

5.5.2.4 Controlled environment agriculture

CEA systems, such as greenhouses and vertical farms, greatly benefit from integrating water and nutrient recycling technologies. In CEA, precise environmental controls – such as temperature, humidity, and light – optimize plant growth. Adding recycled water systems within these environments provides a consistent and efficient water supply and nutrients, essential for maintaining high productivity levels. These systems help reduce water waste and enhance nutrient use efficiency, ensuring that plants receive the optimal resources for growth. By integrating recycling technologies, CEA can operate more sustainably, minimizing its environmental footprint while maximizing crop yields in urban or space-limited settings.

5.5.2.5 Soil amendment

Using nutrient-rich by-products from integrated water and nutrient recycling systems as soil amendments offers significant benefits for sustainable agriculture. These by-products, often derived from anaerobic digestion or constructed wetland processes, can be applied to soil to improve its structure, increase organic carbon content, and boost microbial activity. By enhancing soil health, these amendments contribute to better water retention, reduced erosion, and increased crop productivity. Utilizing recycled nutrients as soil amendments reduces the need for synthetic fertilizers, promoting a more sustainable and environmentally friendly approach to agriculture.

5.5.2.6 Livestock and aquaculture

Integrated water and nutrient recycling systems provide valuable applications in livestock and aquaculture operations. In livestock farming, treated recycled water can be used for animal drinking, cleaning, and cooling, reducing the demand for fresh water and enhancing overall water use efficiency. Nutrient-rich effluents from these systems can grow feed crops or algae, creating a sustainable feed source that reduces reliance on external inputs. In aquaculture, recycled water helps maintain optimal water quality in fish farming systems, supporting the health and growth of aquatic species [14, 15, 125–128].

5.5.3 Economic and environmental benefits

Integrated water and nutrient recycling systems offer substantial economic and environmental advantages. These benefits include cost savings, improved crop yields, and reduced environmental impact, making these systems a valuable component of sustainable agriculture.

5.5.3.1 Economic benefits
Economic benefits include:
- Cost savings: Farmers can reduce reliance on external water supplies and synthetic fertilizers by recycling water and nutrients. This reduction in input costs can lead to significant savings, especially in regions where water and fertilizers are expensive or scarce.

- Increased yields: Improved water and nutrient management can enhance crop productivity and quality. Consistent nutrient delivery through recycled water systems can optimize plant growth, leading to higher yields and profits for farmers.
- Energy efficiency: Many integrated recycling technologies, such as anaerobic digestion, generate renewable energy (biogas) as a by-product. This energy can be used on-site to power agricultural operations, reducing reliance on external energy sources and lowering energy costs.

5.5.3.2 Environmental benefits

Environmental benefits include:
- Resource conservation: Recycling water and nutrients reduces the extraction and consumption of natural resources. This conservation helps protect freshwater ecosystems, reduces the pressure on water supplies, and promotes sustainable resource use.
- Pollution reduction: By capturing and reusing nutrients, integrated systems reduce the risk of nutrient runoff into water bodies, which can cause eutrophication and other environmental problems. Improved nutrient management minimizes the environmental impact of agricultural practices.
- Waste minimization: Integrated systems treat and reuse agricultural and wastewater, reducing the volume of waste that needs to be managed. This waste minimization contributes to a circular economy, where resources are continually reused and recycled.
- Soil health: Nutrient-rich by-products can be used as soil amendments to improve soil structure, increase organic matter content, and enhance microbial activity. Healthy soils are more productive and resilient, supporting sustainable agricultural systems.
- Climate change mitigation: By reducing the need for synthetic fertilizers and minimizing waste, integrated systems lower greenhouse gas emissions associated with fertilizer production and waste management. Renewable energy production from anaerobic digestion reduces the carbon footprint of agricultural operations [4, 14, 15].

5.6 Benefits and challenges of integrated water and nutrient recycling in agriculture

The benefits and challenges of water reuse and recycling in agriculture highlight the significant potential for sustainable water management and the obstacles that must be addressed to optimize these practices.

5.6.1 Environmental, economic, and social benefits of water reuse and recycling in agriculture

Integrating water reuse and recycling practices in agriculture presents a multifaceted approach to sustainability, offering environmental, economic, and social benefits. These practices address the urgent need for water conservation and support broader objectives toward achieving a more sustainable and equitable agricultural sector.

5.6.1.1 Environmental benefits
The environmental benefits of water reuse and water recycling in agriculture include:
–   Reduction in water consumption: Water reuse and recycling significantly reduce the reliance on freshwater sources for agricultural purposes. Treating and reusing wastewater or capturing and recycling rainwater decreases the demand for water extracted from rivers, lakes, and aquifers, helping to preserve these vital natural resources.
–   Pollution reduction: Implementing water reuse and recycling helps minimize the runoff of fertilizers and pesticides into water bodies, reducing eutrophication and the degradation of aquatic ecosystems. Furthermore, by treating wastewater before it is reused or discharged, these practices reduce the load of pollutants entering the environment, leading to cleaner rivers, lakes, and oceans.
–   Resource conservation: Recycling water and nutrients reduces the extraction and consumption of natural resources. This conservation helps protect freshwater ecosystems, reduces the pressure on water supplies, and promotes sustainable resource use.
–   Waste minimization: Integrated systems treat and reuse agricultural and wastewater, reducing the volume of waste that needs to be managed. This waste minimization contributes to a circular economy, where resources are continually reused and recycled.
–   Soil health: Nutrient-rich by-products can be used as soil amendments to improve soil structure, increase organic matter content, and enhance microbial activity. Healthy soils are more productive and resilient, supporting sustainable agricultural systems.
–   Climate change mitigation: By reducing the need for synthetic fertilizers and minimizing waste, integrated systems lower greenhouse gas emissions associated with fertilizer production and waste management. Renewable energy production from processes like anaerobic digestion further reduces the carbon footprint of agricultural operations.

5.6.1.2 Economic benefits

The economic benefits of water reuse and water recycling in agriculture include:

- Cost savings: Farmers who adopt water reuse and recycling technologies can achieve substantial savings on water and energy costs. Reusing water on-site reduces the need for water transportation and treatment, lowering operational expenses. Additionally, using nutrient-rich treated wastewater can decrease the need for chemical fertilizers, further cutting costs.
- Enhanced productivity: Efficient water management through reuse and recycling can lead to more consistent and reliable water supplies for irrigation, enhancing crop yields and quality. This, in turn, can boost the profitability and competitiveness of agricultural enterprises.
- Increased yields: Improved water and nutrient management can enhance crop productivity and quality. Consistent nutrient delivery through recycled water systems can optimize plant growth, leading to higher yields and profits for farmers.
- Energy efficiency: Many integrated recycling technologies, such as anaerobic digestion, generate renewable energy (biogas) as a by-product. This energy can be used on-site to power agricultural operations, reducing reliance on external energy sources and lowering energy costs.

5.6.1.3 Social benefits

The social benefits of water reuse and water recycling in agriculture include:

- Food security: By ensuring a more efficient and sustainable use of water resources, water reuse and recycling contribute to the stability of food production systems. This is particularly crucial in arid and semiarid regions, where water scarcity poses a significant risk to food security.
- Community health and welfare: Improved water management practices help protect water quality that is essential for community health. Additionally, by promoting sustainable agricultural practices, these technologies support rural livelihoods, contributing to social welfare and community resilience.
- Biodiversity conservation: Sustainable water practices in agriculture support the conservation of habitats and biodiversity. By reducing the need for water extraction from natural habitats and minimizing pollution, water reuse and recycling contribute to preserving ecosystems and the species that depend on them [10, 13–15, 129–131].

5.6.2 Overcoming challenges and barriers to water reuse and recycling in agriculture

While water reuse and recycling offer significant benefits for sustainable agriculture, several challenges and barriers hinder their widespread adoption. These challenges

span technical, economic, social, and environmental domains, each requiring strategic solutions:
- *Technical and economic hurdles*:
  - Infrastructure and technology costs: The initial setup for water reuse and recycling systems, including treatment facilities and distribution networks, can be capital-intensive. Smallholder farmers, in particular, may find the costs prohibitive without financial assistance or incentives.
    - Overcoming the barrier: Governments and financial institutions can offer subsidies, low-interest loans, or grants to help offset initial costs. Cooperative models, where farmers share resources and infrastructure, can reduce individual financial burdens.
  - Maintenance and operational complexity: Advanced water treatment technologies require ongoing maintenance and skilled operation. The lack of access to technical expertise and the resources needed for regular system upkeep can impede the effective implementation of water recycling and reuse practices.
    - Overcoming the barrier: Providing training programs for farmers and operators on the maintenance and operation of these technologies is crucial. Establishing partnerships with technology providers for technical support and creating maintenance service networks can also help manage operational complexities.
- *Social acceptability and regulatory issues*:
  - Public perception and acceptance: Concerns about the safety and quality of recycled water can lead to hesitancy among farmers and the public. Misconceptions and lack of awareness about the rigorous treatment processes and safety standards often challenge the acceptability of recycled water in agriculture.
    - Overcoming the barrier: Educational campaigns that provide clear, science-based information about the safety and benefits of recycled water can help change perceptions. Demonstration projects and case studies showcasing the successful and safe use of recycled water can also build confidence among farmers and the public.
  - Regulatory frameworks: Inconsistent or absent water reuse and recycling regulations can create uncertainty and deter adoption. Clear, supportive policies and standards are necessary to guide safe practices and encourage the use of recycled water in agriculture.
    - Overcoming the barrier: Developing comprehensive regulatory frameworks that set straightforward water reuse and recycling standards is essential. Policymakers should collaborate with industry stakeholders to ensure that regulations are practical, encourage adoption, and ensure public safety.

- *Environmental concerns*:
    - Risk of contaminant transfer: Without proper treatment, recycled water may contain residues of pharmaceuticals, chemicals, or pathogens, posing risks to crop health, soil quality, and ultimately, human health. Ensuring that treatment technologies effectively remove or neutralize these contaminants is crucial.
        - Overcoming the barrier: Investing in advanced treatment technologies and continuous monitoring systems can ensure that recycled water meets safety standards. Research and development into more effective and affordable treatment methods can also help mitigate these risks.
    - Long-term soil health: Continually using recycled water, especially if not adequately treated for salinity and other contaminants, can lead to soil degradation over time. Monitoring and managing soil health is essential to mitigate potential adverse effects.
        - Overcoming the barrier: Implementing soil monitoring programs to track changes in soil health and adjusting irrigation practices as needed can help protect long-term soil quality. Using blended water (mixing recycled water with fresh water) and selecting salt-tolerant crops can also reduce the risk of soil degradation [13–15, 120].

Chapter 6
Sustainable food systems advancing water conservation and energy efficiency

Abstract: This chapter explores recent innovations in food systems to improve water conservation and energy efficiency. It examines soil-less farming techniques such as hydroponics and aquaponics, highlighting their potential to reduce water usage in food production. The chapter also covers the development of climate-resilient crops through genetic engineering and selective breeding, which are essential for maintaining agricultural productivity in the face of climate change. Additionally, it discusses strategies for reducing food waste, including advancements in packaging technologies and community-driven initiatives. Integrating renewable energy sources like solar, wind, and biogas into agricultural practices is also explored, focusing on reducing reliance on fossil fuels and minimizing water use. These innovations collectively contribute to a more sustainable and resilient agricultural sector.

6.1 Introduction

The food-water-energy nexus highlights the interconnected relationship between food production, water use, and energy consumption. Efficient management of these resources is essential for sustainable development. Water is critical for crop growth, and energy is necessary for irrigation, processing, and transportation. However, traditional agricultural practices can result in substantial water and energy use, contributing to resource depletion and environmental impacts.

Innovative agricultural approaches solve these challenges by promoting water conservation, optimizing energy use, and reducing food waste. Techniques such as hydroponics, aquaponics, and vertical farming offer ways to reduce water usage and improve efficiency in food production. Additionally, developing climate-resilient crops, implementing advanced technologies, and adopting strategies for food waste reduction are essential for supporting sustainable agricultural practices.

This chapter examines recent innovations in food systems, focusing on their potential to conserve water, enhance energy efficiency, and minimize food waste. It discusses various approaches, including soil-less farming techniques, climate-resilient crops, food waste reduction strategies, and integrating renewable energy into agriculture. The chapter explores these innovations' role in advancing a more sustainable and resilient agricultural sector.

https://doi.org/10.1515/9783111341385-006

6.2 Innovative approaches in food production

As the global population continues to grow, the demand for food increases, putting additional pressure on water and energy resources. Traditional farming methods can lead to high water use and inefficient energy consumption, contributing to environmental challenges such as water scarcity and climate change. Innovative food production approaches are being developed and implemented to address these issues, focusing on maximizing resource efficiency, reducing waste, and enhancing sustainability. The agricultural sector can reduce its environmental impact by adopting advanced technologies and practices while supporting food security. This section explores various innovative food production methods, including hydroponics, aquaponics, and vertical farming, and examines their potential to transform food production.

6.2.1 Hydroponics and aquaponics

Hydroponics and aquaponics are innovative agricultural systems that offer efficient alternatives to traditional soil-based farming. Hydroponics involves growing plants in a nutrient-rich water solution, allowing precise control over nutrient delivery and reducing water usage compared to conventional methods. Aquaponics combines hydroponics with aquaculture, where fish waste provides nutrients for plant growth, and plants help filter and clean the water, creating a closed-loop system. These approaches conserve water and maximize resource efficiency, making them valuable for sustainable food production. This section examines the principles and applications of hydroponics and aquaponics, highlighting their potential to contribute to more sustainable agricultural practices.

6.2.1.1 Explanation of soil-less farming techniques

Hydroponics is a method of growing plants without soil using nutrient-rich water solutions to deliver essential minerals directly to the plant roots. This technique allows for precise control over the nutrient levels, resulting in faster plant growth and higher yields. Hydroponic systems can be set up in various configurations, such as nutrient film technique, deep water culture, and aeroponics, each tailored to specific plant needs. In contrast, aquaponics combines hydroponics with aquaculture, the practice of raising fish, to create a symbiotic environment where plants and fish benefit from each other's presence. In an aquaponic system, fish waste provides an organic nutrient source for the plants, while the plants help to filter and clean the water for the fish. These systems mimic natural ecosystems and promote sustainability by recycling nutrients and water [132–134].

6.2.1.2 Benefits of water recirculation in self-contained systems

One of the most significant advantages of hydroponics and aquaponics is their efficient water use. Traditional soil-based agriculture can lose up to 70% of its water to evaporation and runoff, whereas hydroponic systems recirculate water, reducing overall consumption by up to 90%. This water efficiency makes hydroponics an ideal solution for regions facing water scarcity. Similarly, aquaponic systems continuously recirculate water between the fish tanks and the plant beds. The plants absorb the nutrients from the fish waste, effectively purifying the water, which is then cycled back to the fish tanks. This closed-loop system drastically reduces water usage and minimizes waste. Additionally, these systems eliminate the need for chemical fertilizers and pesticides, further contributing to environmental sustainability. The combination of water efficiency and nutrient recycling makes hydroponics and aquaponics highly sustainable farming methods that can significantly reduce the environmental footprint of food production [135, 136].

6.2.1.3 Integration into urban agricultural initiatives

Both hydroponics and aquaponics are increasingly being integrated into urban agricultural initiatives, addressing the growing need for local, sustainable food production in cities. Urban environments often face limited arable land, high population density, and environmental pollution. Hydroponic and aquaponic systems can be established where traditional farming is not feasible by utilizing rooftops, vacant lots, and indoor spaces. These systems can contribute to urban food security, reduce the carbon footprint associated with food transportation, and create green spaces that improve air quality and community well-being. Urban farms can provide fresh, locally grown produce to city dwellers, reduce the reliance on imported food, and foster community through urban gardening projects. The scalability and adaptability of hydroponics and aquaponics make them ideal for urban settings, where they can play a crucial role in building sustainable and resilient food systems [4].

Case 6.1 Bustanica's sustainable approach to innovative hydroponic farming in Dubai

Bustanica, a leading vertical farm in Dubai, uses hydroponic technology to grow various leafy greens. This advanced 330,000 ft^2 facility produces over 1,000,000 kg of high-quality greens annually while reducing water usage by 95% compared to traditional farming. The farm leverages advanced technologies, including machine learning and artificial intelligence, and is operated by a team of experts in agronomy, engineering, horticulture, and plant science. Bustanica's continuous production ensures that the greens are always fresh, pesticide-free, and chemical-free. Bustanica's closed-loop system enhances water efficiency by recirculating it through the plants and capturing evaporated water for reuse, saving approximately 250 million l of water annually compared to traditional farming methods. This innovative approach does not negatively impact soil quality and significantly benefits the world's endangered soil resources.

Additionally, Bustanica's year-round production is unaffected by weather or pests, ensuring a consistent supply of fresh, uncontaminated greens. The greens are packaged to avoid the need for washing, preventing potential damage and contamination [137].

6.2.2 Vertical farming

Vertical farming is an innovative agricultural method that involves growing crops in vertically stacked layers, often in controlled indoor environments. This technique significantly reduces water and land use, making it a sustainable alternative to traditional farming practices.

6.2.2.1 Role of vertical farming in reducing water and land use

Vertical farming significantly reduces water and land use using advanced technologies and efficient growing techniques. Traditional agriculture often requires large tracts of arable land and extensive water resources to cultivate crops. In contrast, vertical farming maximizes space by growing crops in stacked layers, which can be housed in buildings, shipping containers, or specially designed structures. This vertical arrangement allows for higher crop yields per square foot of land, making it possible to grow more food in less space.

Moreover, vertical farming employs hydroponic, aeroponic, and aquaponic systems that use water more efficiently than traditional soil-based farming. These systems deliver nutrients directly to plant roots through nutrient-rich water solutions, significantly reducing water wastage. Water in these systems is recirculated, meaning excess water is collected, filtered, and reused, minimizing the overall water consumption. Studies have shown that vertical farming can reduce water usage by up to 95% compared to conventional farming methods. This high efficiency in water use makes vertical farming particularly suitable for regions facing water scarcity [138–140].

6.2.2.2 Benefits of vertical farming in urban areas for reducing food transport carbon footprint

Vertical farming is particularly beneficial in urban settings, where it can help reduce the carbon footprint associated with food transport. Urban areas often rely on food transported from rural farms over long distances. This transportation process contributes to greenhouse gas emissions and increases the overall carbon footprint of the food supply chain. By implementing vertical farms within or near urban centers, cities can produce fresh, locally grown food, significantly reducing the need for long-distance transportation.

The proximity of vertical farms to urban consumers also ensures that produce is fresher and has a longer shelf life, as it spends less time in transit. This not only improves food quality but also reduces food waste, which is a significant issue in the traditional supply chain. Additionally, vertical farming can operate year-round, regardless of external weather conditions, ensuring a consistent supply of fresh produce and enhancing food security in urban areas.

Urban vertical farms can be established in various locations such as on rooftops, abandoned warehouses, or purpose-built structures. These farms can transform unused or underutilized urban spaces into productive agricultural sites, contributing to the greening of cities and improving urban biodiversity. By integrating vertical farming into urban landscapes, cities can take a significant step toward sustainability, reducing their environmental impact and creating a more resilient food system [140, 141].

Case 6.2 GigaFarm's impact on United Arab Emirates' food production
The GigaFarm project, part of the United Arab Emirates' Food Tech Valley initiative, is poised to transform sustainable food production in the region. Managed by ReFarm, the 83,612 m^2 facility aims to produce 3 million kg of food annually using vertical farming technologies. The project focuses on addressing the environmental impact of food production, including reducing greenhouse gas emissions from long-distance transportation and decreasing reliance on traditional agricultural practices. GigaFarm employs innovative waste-to-value technologies, recycling food scraps and sewage into compost, animal feed, clean water, and energy. This circular approach minimizes waste and enhances sustainability. GigaFarm significantly reduces water usage and fertilizer consumption using controlled environments, making food production more efficient. The facility's proximity to urban centers allows for fresher produce and reduces the carbon footprint associated with food imports. To replace up to 1% of the UAE's food imports, GigaFarm is expected to play a crucial role in enhancing food security in the region. The project also highlights the potential of vertical farming to address the challenges of water scarcity, limited arable land, and climate change. By integrating modern technology with sustainable practices, GigaFarm exemplifies a forward-thinking approach to agriculture that could serve as a model for other regions facing similar challenges [142].

6.3 Climate-resilient crops

Climate-resilient crops are developed to withstand changing climate conditions such as increased temperatures, drought, and extreme weather events. These crops are enhanced through selective breeding and genetic engineering to maintain productivity and ensure food security in the face of climate variability. This section explores the development and benefits of climate-resilient crops, focusing on their role in supporting sustainable agriculture and adapting to the impacts of climate change.

6.3.1 Development of drought-resistant and climate-resilient crop varieties

Developing drought-resistant and climate-resilient crop varieties is essential for sustaining agricultural productivity in regions affected by climate change. Researchers use advanced breeding techniques and genetic engineering to create crops that thrive under challenging conditions such as prolonged drought, heat stress, and unpredictable weather patterns. These crop varieties are designed to use water more efficiently, maintain yields in adverse environments, and contribute to the overall resilience of farming systems.

6.3.1.1 Current challenges in agriculture

Climate change significantly impacts crop yields and agricultural productivity. Increasingly, farmers face issues such as prolonged droughts, extreme temperatures, and unpredictable weather patterns, all of which contribute to lower yields and increased crop failure. These challenges threaten food security and the livelihoods of those dependent on agriculture. Traditional crop varieties often cannot cope with the stress caused by these adverse conditions, leading to a pressing need for developing crops that can survive and thrive despite environmental extremes. Enhancing the resilience of crops through innovative breeding and genetic engineering techniques is essential to ensure stable and productive agricultural systems in the face of ongoing climate change [143–147].

6.3.1.2 Examples of drought-resistant crops

Drought-resistant crops are crucial in addressing water scarcity and maintaining agricultural productivity in arid and semiarid regions. Key examples of such crops include drought-tolerant maize, sorghum, and millet. These crops have been specifically developed or selected for their ability to withstand prolonged periods of low water availability.

Drought-tolerant maize is engineered to perform well under water-stressed conditions, featuring traits such as efficient water use and deep root systems that allow the plant to access moisture from deeper soil layers. Naturally adapted to hot and dry environments, sorghum exhibits high drought resistance due to its ability to reduce water loss through smaller leaf areas and waxy leaf surfaces. Millet, another staple in dry regions, thrives with minimal water and can grow in poor soil conditions, making it an ideal crop for areas with low rainfall.

The importance of these drought-resistant crops cannot be overstated in regions prone to drought and water scarcity. They provide a reliable source of food and income for farmers, ensuring food security and supporting livelihoods even under adverse climatic conditions. By cultivating drought-resistant crops, farmers can sustain agricultural productivity and mitigate the impacts of climate change on food production [148–151].

6.3.2 Techniques in crop genetic engineering and selective breeding

Genetic engineering involves directly modifying an organism's DNA to enhance specific traits, while selective breeding involves choosing parent plants with desirable characteristics to produce offspring with improved traits. Both techniques are vital for developing climate-resilient crops, enabling them to withstand extreme weather conditions and ensure food security.

6.3.2.1 Genetic engineering techniques

Genetic engineering involves altering the DNA of crops to introduce or enhance specific traits, often using advanced tools like Clustered Regularly Interspaced Short Palindromic Repeats (CRISPR) and transgenic methods. CRISPR allows precise editing of an organism's genome, making it possible to introduce desirable traits quickly and accurately. Transgenic crops involve inserting genes from other species to confer beneficial characteristics.

Examples of genetically engineered crops with enhanced climate resilience include drought-tolerant wheat and flood-resistant rice. Drought-tolerant wheat is modified to use water more efficiently and maintain productivity under water-stressed conditions. Flood-resistant rice has been engineered to survive prolonged submersion, a common issue in flood-prone areas.

Genetically engineered crops have significant benefits. They can increase yields, reduce reliance on chemical inputs, and improve resistance to pests and diseases. These crops can also be tailored to thrive in specific environmental conditions, making them essential for addressing the challenges posed by climate change.

However, genetically engineered crops have potential risks. Concerns include the potential for unintended environmental impacts, such as crossbreeding with wild relatives and developing resistance to pests. Additionally, there are ethical and socioeconomic considerations, such as large corporations' control of seed patents and the public acceptance of genetically modified organisms. Balancing these benefits and risks is crucial for developing and deploying genetically engineered crops [145, 152–155].

6.3.2.2 Selective breeding techniques

Selective breeding involves choosing parent plants with desirable traits to produce offspring that exhibit those traits more strongly. This traditional method has been used for centuries to enhance crop resilience and productivity. By repeatedly selecting and breeding plants that show tolerance to specific stresses, farmers and scientists have developed varieties better suited to challenging environmental conditions.

Crops improved through selective breeding for climate resilience include heat-tolerant tomato varieties and salt-tolerant barley. Heat-tolerant tomatoes can thrive in higher temperatures, ensuring yield stability in hotter climates. Salt-tolerant barley can grow in soils with higher salinity, making it suitable for areas affected by soil salinization.

The advantages of selective breeding include its long history of success and acceptance as well as the ability to improve crops without introducing foreign genes. However, selective breeding is often slower and less precise than genetic engineering. It relies on existing genetic variation within a species and can take many generations to achieve desired results. Despite these limitations, selective breeding remains a valuable tool for developing climate-resilient crops [156, 157].

Case 6.3 Ancient Environmental Genomics Initiative enhancing modern agriculture for food security

The Ancient Environmental Genomics Initiative for Sustainability (AEGIS) aims to address global food security challenges by leveraging the genetic diversity of ancient environmental DNA (eDNA). With £66 million in funding from the Novo Nordisk Foundation and Wellcome, AEGIS seeks to understand how ancient plants adapted to historical climate changes. By studying genetic material preserved in natural sources like soil, ice, and water, scientists can develop strategies to make modern crops more resilient to current climate challenges. Modern agricultural practices have reduced genetic diversity in crops, making them vulnerable to extreme weather conditions. AEGIS brings together experts from institutions such as EMBL-EBI, the University of Copenhagen, and the University of Cambridge to utilize advanced DNA sequencing and bioinformatics tools. The program's findings will be publicly available, supporting global efforts to enhance food security. This research aims to create climate-resilient crops and promote sustainable agricultural practices for the future [158].

6.4 Food waste reduction strategies

Food waste is a significant global issue that impacts food security, environmental sustainability, and economic efficiency. Reducing food waste is essential to optimize resource use and ensure more food reaches needy people. This section explores various strategies to minimize food waste, focusing on innovative packaging technologies, community-driven initiatives, and policy reforms and regulations. By adopting these strategies, we can significantly reduce the amount of food lost or wasted throughout the supply chain, from production to consumption, thereby promoting a more sustainable and efficient food system.

6.4.1 Technology-driven solutions

Innovative packaging technologies are crucial in reducing food waste by extending shelf life and improving product preservation.

6.4.1.1 Biodegradable materials

Biodegradable packaging materials are designed to break down naturally in the environment, reducing the accumulation of plastic waste. These materials offer several benefits including minimizing environmental pollution and reducing the carbon footprint associated with traditional plastic packaging. Common biodegradable materials used in food packaging include polylactic acid (PLA), derived from renewable resources like corn starch, and starch-based materials, which decompose more quickly than conventional plastics.

Biodegradable packaging helps reduce plastic waste and supports a circular economy by promoting renewable resources. As these materials break down, they return to the ecosystem without leaving harmful residues, making them an environmentally friendly alternative. By incorporating biodegradable packaging, the food industry can significantly decrease its reliance on fossil-fuel-based plastics, contributing to a cleaner and more sustainable environment [159, 160].

6.4.1.2 Water-vapor-permeable materials

Water-vapor-permeable packaging materials allow the controlled exchange of water vapor between the packaged product and the external environment. This permeability helps maintain optimal humidity levels within the package, which is crucial for preserving the freshness and quality of perishable products. Examples of such materials and technologies include breathable films and modified atmosphere packaging (MAP). Breathable films are designed to balance moisture and gas exchange, while MAP adjusts the composition of gases inside the package to slow spoilage.

These water-vapor-permeable materials significantly extend the shelf life of fruits, vegetables, and other perishable items by preventing moisture build-up that can lead to mold growth and spoilage. By maintaining the correct humidity levels, these packaging solutions help keep products fresher for longer, reducing food waste and enhancing food safety. The extended shelf life provided by these materials benefits consumers and retailers by reducing the frequency of product turnover and spoilage-related losses [161].

6.4.1.3 Impact on product shelf life and food wastage reduction

Innovative packaging is vital in maintaining food quality and safety, directly impacting waste reduction. Advanced materials and technologies, such as biodegradable and water-vapor-permeable packaging, can significantly extend the shelf life of perishable products. These packaging solutions help maintain optimal conditions within the package, preserving freshness, texture, and nutritional value.

Extended shelf life reduces food spoilage by slowing down the natural degradation processes that lead to the loss of food quality. When products stay fresh for extended periods, they are less likely to be discarded due to spoilage. This benefits consumers by providing high-quality, longer-lasting food and helps retailers reduce losses associated with unsellable goods. Additionally, longer shelf life allows for better inventory management and more efficient distribution, minimizing the chances of products expiring before they reach consumers.

Innovative packaging contributes to a more sustainable food system by reducing spoilage and waste. It ensures that more food produced is consumed, thus optimizing resource use and decreasing the environmental impact associated with food production, transportation, and disposal. This approach aligns with broader efforts to enhance food security and sustainability globally [162].

6.4.2 Community and policy initiatives

Community and policy initiatives are crucial in reducing food waste and promoting sustainable agricultural practices. Grassroots efforts, such as food-sharing platforms and community education programs, empower individuals and local groups to actively participate in minimizing food wastage. At the same time, effective policies and regulations are essential for creating an enabling environment that supports these community-driven efforts. Policymakers can significantly impact food waste management by implementing straightforward food labeling standards, incentivizing food donation, and promoting waste reduction practices. This section explores the various community initiatives and policy reforms that collectively contribute to reducing food waste and fostering a more sustainable food system.

6.4.2.1 Food-sharing platforms

Food-sharing platforms are online or app-based services facilitating surplus food sharing within communities. These platforms allow individuals and businesses to donate excess food that might otherwise go to waste, connecting them with those who need it.

The benefits of food-sharing platforms are significant. They help reduce household and community-level food waste by ensuring that excess food is redistributed rather than discarded. This supports those in need and promotes a culture of sharing and sustainability. By reducing food waste, these platforms contribute to environmental conservation and resource efficiency, making them an essential tool in the fight against food waste.

6.4.2.2 Community education and awareness programs

Community education and awareness programs are vital for reducing food waste. Educating communities about the importance of food waste reduction helps individuals understand the environmental, economic, and social impacts of their food disposal habits. These programs often include workshops, campaigns, and school initiatives that teach practical ways to minimize waste such as proper food storage, meal planning, and understanding food labels.

Community engagement plays a crucial role in promoting sustainable food practices. Communities are more likely to adopt and advocate for waste reduction practices when actively involved. Engaged communities can foster a culture of sustainability, where individuals collectively work toward reducing food waste. These programs help create lasting change by empowering people with the knowledge and skills to make more sustainable choices [4].

6.4.3 Policy reforms and regulations

Policy reforms and regulations are essential components in the effort to reduce food waste and promote sustainable food systems. Effective policies create a supportive framework encouraging individuals and businesses to adopt waste-reduction practices. Governments can address systemic issues contributing to food waste by implementing clear and comprehensive regulations and incentivizing sustainable behaviors. This section examines the role of policy reforms in enhancing food labeling standards, promoting food donation, and encouraging resource-efficient practices within the food industry. Through targeted policy measures, significant progress can be made toward minimizing food waste and ensuring a more sustainable and equitable food supply chain.

6.4.3.1 Enhancing food labeling standards

Clear and accurate food labeling is crucial for reducing food waste and ensuring consumers make informed choices. Misunderstanding food labels often leads to prematurely disposing of food, contributing significantly to household food waste. Accurate labeling can help consumers better understand the shelf life of products, thereby reducing unnecessary waste.

A critical reform in food labeling is distinguishing between "best before" and "use by" dates. "Best before" dates indicate when a product will be at its best quality, while "use by" dates are related to safety and indicate when a product should no longer be consumed. Clarifying these terms can prevent confusion and encourage consumers to use their judgment about the edibility of food beyond the "best before" date, as many products remain safe to eat.

These labeling reforms can substantially impact consumer behavior. By providing clear information, consumers are more likely to use food before it spoils and less likely to discard items prematurely. This change can lead to a significant reduction in food waste at the household level. Additionally, better labeling helps retailers manage their inventory more effectively, reducing the amount of unsold food thrown away.

Enhancing food labeling standards is a simple yet effective strategy to combat food waste. By improving the clarity and accuracy of food labels, consumers can make better decisions, ultimately contributing to a more sustainable and efficient food system [4].

6.4.3.2 Government and institutional policies

Government and institutional policies are critical in supporting food waste reduction initiatives. By establishing a regulatory framework and providing incentives, governments can encourage individuals and businesses to adopt practices that minimize food waste. Effective policies address the root causes of food waste and promote sustainable resource management.

One example of a successful policy is the implementation of food donation laws. These laws protect food donors from liability and encourage restaurants, supermarkets, and other food businesses to donate surplus food to charities and food banks rather than discarding it. Countries like the United States and France have enacted such laws, significantly increasing food donations and reducing food waste.

Another effective policy approach is providing incentives for food waste reduction. Governments can offer tax breaks or grants to businesses implementing waste reduction strategies such as composting, recycling, or investing in food waste reduction technologies. These incentives help offset the costs of adopting new practices and encourage broader participation.

The impact of these policies on large-scale food waste reduction and resource conservation is substantial. By reducing the amount of food waste generated, these policies help conserve resources such as water, energy, and labor that go into food production. Additionally, less food waste means reduced greenhouse gas emissions from landfills, contributing to environmental sustainability.

Overall, government and institutional policies are essential for driving large-scale change in food waste reduction. Governments can create an environment that supports sustainable practices and resource conservation through protective legislation, financial incentives, and comprehensive regulations [4].

Case 6.4 Love Food Hate Waste NZ reducing food waste for a sustainable future
Love Food Hate Waste NZ is the leading authority in New Zealand on reducing food waste. Through research and expertise, the organization conducts engaging campaigns that offer practical ways to minimize food waste, helping Kiwi households save time and money while benefiting the environment. Supported by WasteMINZ, 52

councils, and funding from the Ministry of Environment, Love Food Hate Waste NZ is dedicated to making a positive impact. The mission of Love Food Hate Waste NZ is to inspire and enable New Zealanders to waste less food. By changing shopping, cooking, and consumption habits, individuals can save money, reduce their carbon footprint, and promote sustainability. The organization emphasizes that real change starts in the kitchen by addressing food waste at its source. Citizens are encouraged to plan their shopping, check their fridges and pantries, and plan meals. Additionally, best storage practices are promoted such as refrigerating food within two hours, consuming it within two days, and freezing it for up to two months. This comprehensive approach aims to create a sustainable future by reducing food waste [163].

6.5 The water-energy-food nexus: enhancing efficiency

The water-energy-food nexus highlights the interdependence between water, energy, and food production, emphasizing the need for efficient resource management to achieve sustainability. Energy is crucial at every stage of the food supply chain, from irrigation to transportation. Traditional agricultural practices often lead to high energy consumption, increasing greenhouse gas emissions, production costs, and water demand. This section explores innovative approaches and technologies to enhance energy efficiency in food production such as integrating renewable energy sources and advanced farming techniques. The agricultural sector can reduce its environmental footprint, lower costs, and contribute to a more sustainable and resilient food system by optimizing energy use.

6.5.1 Renewable energy in agricultural practices

Integrating renewable energy sources into agricultural practices is crucial for enhancing sustainability and reducing environmental impacts. Renewable energy, such as solar, wind, and biogas, offers clean alternatives to traditional fossil fuels, helping to lower greenhouse gas emissions and reduce energy costs. This section explores how these renewable energy sources are being utilized in agriculture to create more efficient and sustainable food production systems.

6.5.1.1 Solar energy
Solar energy is a robust and sustainable resource for agriculture. By harnessing sunlight, farmers can generate electricity and heat, reducing their reliance on fossil fuels. Solar-powered irrigation systems, for example, use photovoltaic panels to pump water, ensuring a reliable water supply without significant energy costs. Solar green-

houses utilize solar panels to maintain optimal growing conditions, extending the growing season and improving crop yields. Additionally, solar dryers efficiently dehydrate produce, reducing spoilage and extending shelf life.

The benefits of using solar energy in agriculture are substantial. It helps reduce greenhouse gas emissions by decreasing the need for fossil fuel-based energy sources. Moreover, solar energy systems can significantly lower energy costs for farmers, making agricultural operations more economically viable. Adopting solar technology in agriculture supports environmental sustainability and economic resilience [4].

6.5.1.2 Wind energy

Wind energy is another renewable resource increasingly used in agricultural settings. Small-scale wind turbines can be installed on farms to generate electricity, providing a clean and cost-effective power source. These turbines can power various farm operations, from lighting and ventilation systems to water pumps and electric fences.

The benefits of wind energy in agriculture include enhanced sustainability and reduced energy costs. Wind turbines produce no emissions, contributing to a cleaner environment. Additionally, harnessing wind energy helps farmers become more energy-independent and resilient to fluctuating energy prices, ultimately supporting the long-term sustainability of agricultural practices [4].

6.5.1.3 Biogas

Biogas production from agricultural waste is an efficient way to convert organic matter into usable energy. Microorganisms break down animal manure, crop residues, and other organic waste through anaerobic digestion, producing biogas and nutrient-rich digestate. Biogas can be used for heating, electricity generation, and as a fuel for farm machinery, providing a versatile energy source.

The environmental and economic advantages of biogas systems are notable. They help reduce greenhouse gas emissions by capturing methane that would otherwise be released into the atmosphere. Additionally, biogas systems promote waste management and provide a renewable energy source, contributing to environmental sustainability and economic efficiency in agricultural operations [4].

6.5.1.4 Reduction of the water footprint associated with traditional energy production methods

The use of renewable energy sources significantly reduces water usage in agriculture. Traditional energy production methods, such as coal and natural gas power plants, require substantial water for cooling and processing. In contrast, renewable energy systems like solar and wind use minimal water during operation. For instance, solar panels convert sunlight into electricity without needing water for cooling, and wind turbines generate power from wind without any water consumption. This significant difference in water usage makes renewable energy a much more sustainable option. Agriculture can decrease its water footprint by adopting renewable energy technologies, conserving this vital resource and promoting more sustainable practices [4].

Case 6.5 Green energy transition for sustainable agriculture in Botswana
In 2022, Botswana's National Development Bank (NDB) launched a funding program for small and medium-sized off-grid farmers and horticultural producers. The program seeks to reduce greenhouse gas emissions and enhance climate adaptation by promoting renewable energy solutions and water-efficient irrigation practices. This initiative, supported by €850,000 in funding from the International Climate Initiative, addresses Botswana's arid climate and its vulnerability to extreme weather events, such as droughts and floods, which have increased due to climate change. Historically, off-grid farmers in Botswana have relied on diesel generators, which have significant environmental and economic costs. The NDB's funding scheme supports replacing diesel engines with solar energy, installing solar water pumps, and adopting energy-saving solutions for cooling and lighting. In addition, the program promotes water-efficient irrigation techniques, such as drip irrigation, and controlled farming environments like shade netting and hydroponics to improve water and energy efficiency. The program focuses on empowering women and youth, with 25% of the funds allocated to women-led and youth-led enterprises. By reducing reliance on diesel and enhancing sustainable water use, the initiative aims to improve Botswana's agricultural productivity while mitigating the environmental impact of farming in off-grid areas. Through this project, the NDB fosters sustainable agricultural practices that enhance production and contribute to Botswana's climate resilience [164].

6.5.2 Energy recovery from food waste

Energy recovery from food waste is a sustainable approach that converts organic waste into valuable energy resources, reducing waste and supporting environmental sustainability. Technologies for energy recovery from food waste offer sustainable solutions to reduce waste and generate renewable energy.

6.5.2.1 Anaerobic digesters
Anaerobic digestion is a biological process that breaks down organic matter, such as food waste, without oxygen. This process occurs in an anaerobic digester, where microorganisms decompose the waste, producing biogas and nutrient-rich digestate. Biogas, primarily composed of methane and carbon dioxide, can be used as a renewable energy source for heating, electricity generation, and vehicle fuel. Digestate, the residual material, can be used as a high-quality fertilizer.

The benefits of using anaerobic digesters to convert food waste into biogas and digestate are significant. They reduce the volume of waste sent to landfills, lower greenhouse gas emissions, and produce renewable energy. Additionally, the digestate by-product enriches the soil, enhancing agricultural productivity. Examples of anaerobic digester systems are found on farms, where manure and crop residues are processed, and in food processing facilities, where organic waste from production lines is converted into energy and fertilizer [4].

6.5.2.2 Conversion of food waste into energy

Several methods exist for converting food waste into energy including biogas production and composting with energy capture. In biogas production, anaerobic digesters break down organic waste to produce biogas, which can be used for various energy needs. Composting with energy capture involves decomposing organic matter to generate heat, which can be harnessed for energy use.

The benefits of these methods are multifaceted. They help reduce the amount of waste sent to landfills, mitigating environmental pollution and greenhouse gas emissions. Additionally, these processes generate renewable energy, contributing to the reduction of reliance on fossil fuels. The by-products of these methods, such as compost and digestate, are valuable for agricultural use, improving soil health and fertility. These technologies promote sustainability and resource efficiency in agriculture and food processing by converting food waste into energy [4].

6.5.2.3 Role in water conservation

Energy recovery from food waste plays a crucial role in water conservation by reducing the need for water-intensive waste disposal methods. Traditional waste management practices, such as landfilling and incineration, often require significant amounts of water for processing and maintaining hygiene. In contrast, anaerobic digesters and other energy recovery systems use minimal water, offering a more sustainable alternative.

Anaerobic digesters convert food waste into biogas and digestate without extensive water usage. This process generates renewable energy and produces a nutrient-rich by-product that can be used as fertilizer, reducing the need for synthetic fertilizers, which are water-intensive. Additionally, the liquid fraction of the digestate can be used for irrigation, further conserving water resources.

Other energy recovery methods, such as composting with energy capture, also have water-saving benefits. Composting reduces organic waste volume, decreasing water needs in waste management processes. The heat generated during composting can be harnessed for energy, efficiently using resources without additional water input.

The impact of these systems on overall water usage in agricultural and food processing operations is significant. By diverting food waste from traditional disposal methods to energy recovery systems, water usage is reduced at multiple stages of the waste management process. Water conservation supports more sustainable agricultural practices and helps address the broader issue of water scarcity, making energy recovery from food waste a vital component of resource-efficient food production systems [4].

Case 6.6 Rialto Bioenergy Facility receives grant to boost organic waste recycling
Rialto Bioenergy Facility, LLC (Rialto) received a $4 million grant from the Organics Grants program to fund the installation of an anaerobic digester and a freezer to salvage food that would otherwise go to landfill. These installations are crucial for the operational success of Rialto's new Southern California facility, which is expected to recycle 300,000 tons of organic waste annually. The facility will house a 3 million gallon anaerobic digester capable of converting 300 tons of food waste into 4.2 MW of renewable biopower daily. Approximately one-fourth of this energy will be used on-site, with the remainder exported to the power grid. Additionally, the project will produce 30,660 tons of marketable dry fertilizer pellets annually from the solid residuals of the anaerobic digestion process, preventing the emission of 22,630 metric tons of carbon dioxide equivalent per year by offsetting fossil fuel use. A new 90-ton capacity freezer in Redlands will support Helping Hands Pantry, a food rescue partner of Rialto, in feeding food-insecure residents. This initiative will help rescue up to 3,276 tons of frozen food annually and create 14 new full-time, permanent jobs for disadvantaged and low-income communities in the area [165].

Chapter 7
Urban agriculture and water management circular solutions for urban landscapes

Abstract: This chapter examines urban agriculture's role in circular urban ecosystems, highlighting its contributions to food security, reducing urban heat islands, and improving water management. Urban agriculture enhances environmental stewardship and resource efficiency by integrating food production into city environments. The chapter covers various types of urban agriculture, from downtown areas to peri-urban regions, and their benefits, such as enhancing green spaces, improving food accessibility, and fostering community building. It discusses urban agriculture's integration into the circular economy, addressing environmental, social, and economic challenges. Additionally, it explores closed-loop water systems and waste-to-resource initiatives, which transform organic waste into compost and bioenergy. The chapter underscores urban agriculture as a nature-based solution, providing ecosystem services that enhance urban resilience and sustainability. Finally, it outlines strategies for promoting urban agriculture including favorable policies, financial support, and community engagement.

7.1 Introduction

Urban agriculture is a transformative component of circular urban ecosystems, enhancing food security, reducing the urban heat island effect, and improving water management. Integrating food production within city environments promotes long-term environmental stewardship and resource efficiency. Historically rooted in Roman peri-urban farms and medieval monastic gardens, urban agriculture now includes community and rooftop gardens, beekeeping, aquaculture, vertical farming, hydroponics, and aquaponics. It also features direct food distribution through farmers' markets. Urban agriculture supports local economies, strengthens social ties, and improves ecological health. Adopting circular economy principles shortens supply chains, ensures faster delivery, and improves food quality and security.

This chapter covers various types of urban agriculture, from densely built downtown areas to peri-urban regions. These initiatives enhance green spaces, improve food accessibility, promote food literacy, provide job skills, create employment, and foster community building. The chapter examines urban agriculture's role in the circular economy, addressing environmental, social, and economic challenges and creating sustainable urban ecosystems. It discusses closed-loop water systems and waste-to-resource initiatives and highlights urban agriculture as a nature-based solution. Finally, it outlines strategies for promoting urban agriculture including favorable policies, financial support, and community engagement.

https://doi.org/10.1515/9783111341385-007

7.2 Types of urban agriculture

Urban agriculture occurs in various urban settings from densely built downtown areas to the open spaces of periurban regions. These initiatives fulfill multiple urban functions, including enhancing green spaces, ensuring food security, improving food accessibility, promoting food literacy, providing job skills training, creating employment opportunities, and fostering community building.

Urban agriculture comes in various shapes and sizes:

- *Allotment gardens*: These are areas divided into small plots that individuals can rent under a tenancy agreement. These gardens often originate from municipal initiatives on public land and are typically governed by formal regulations, sometimes adhering to specific regional or national laws. Management of allotment gardens may be overseen by an organized group or established as an allotment garden association, requiring participants to become members of the organization.
- *Backyard gardens*: Backyard gardens involve cultivating food on private property. The harvested produce is often shared with friends, family, and neighbors, resulting in a surplus. Additionally, the food can be stored and preserved. These gardens benefit communities by allowing neighbors to collaborate and use various farming techniques, which can improve overall yields.
- *Community gardens*: Community gardens are grassroots initiatives maintained by residents. They focus on vegetable growing, social networking, and community building. Typically small and located in urban spaces, these gardens emphasize organic production and often include composting facilities. They also host educational and cultural activities. Community-established rules guide them, and agreements with authorities or property owners are usually negotiated but not always formalized.
- *Commercial farms*: Commercial farms in urban areas, like rural farms, are operated by for-profit organizations. These farms often specialize in niche produce and utilize high-efficiency methods such as vertical and soilless farming.
- *Educational gardens*: Educational gardens serve as teaching tools that focus on the production, processing, and consumption of food, highlighting their environmental impact. These gardens raise public awareness and promote eco-friendly gardening practices. Found in schools, kindergartens, and environmental centers, they provide communities with garden-based learning and educational services. The success of educational gardens often depends on dedicated teachers and public support, and municipal policies can support their creation.
- *Forest gardening*: Forest gardening involves cultivating gardens within urban forests. This practice includes growing crops, vegetables, and fruits in city environments. Forests provide a favorable environment for crop development, helping to protect urban forests and mitigate deforestation. Additionally, forest gardening

supports afforestation efforts, promoting tree planting as a measure against global warming in urban areas.

- *Greenhouses*: This involves practicing agriculture in residential, commercial, and communal urban spaces. Depending on the crops being grown, they require significant land. Greenhouses allow farmers to cultivate crops year-round by providing a controlled environment with specific conditions needed for plant growth.
- *Green walls*: Green walls involve growing vegetation or food crops on the exterior or interior surfaces of walls. They are space-efficient and use mechanisms to supply water and soil integrated into the wall structure. Green walls effectively reduce stormwater runoff.
- *Hydroponics*: Hydroponic and aeroponic systems represent one of the most promising water-efficient approaches to urban agriculture. These systems consume considerably less water than conventional farming methods, as plants are cultivated in nutrient-rich water rather than soil. Hydroponic systems can reduce water usage by up to 90% while requiring fewer pesticides and herbicides, resulting in a healthier and more sustainable food source.
- *Institutional farms and gardens*: Institutional farms and gardens, associated with institutions such as churches, hospitals, schools, and prisons, serve to enrich and educate their members. Managed by a paid overseer, these farms and gardens ensure their produce is well-utilized. Operating on institutional property often reduces infrastructure-related challenges.
- *Micro-farming*: Micro-farming in and around the house is a widespread urban agriculture practice in many cities. It typically involves growing vegetables, herbs, and medicinal plants in small areas within homes (such as balconies, windowsills, and rooftops) and around them (like front and backyards). Small animals, such as rabbits and chickens, may also be kept. This low-investment activity attracts diverse participants, from low-income to high-income households, motivated by subsistence, leisure, environmental awareness, or a desire to grow their food. Surpluses are often shared, bartered, or occasionally sold.
- *Multifunctional farms*: Multifunctional farms integrate various activities beyond traditional food production including offering fresh food, training, recreational activities, educational programs, and health services. They also incorporate water and landscape management in and around urban areas. Practiced by small- and large-scale farmers and urban investors, these farms diversify their operations to create new income streams, reduce costs, and meet urban demand for leisure and environmental services.
- *Rooftop gardens*: Rooftop gardens utilize available rooftop spaces in urban areas to grow vegetables, fruits, and herbs. They help mitigate urban heat islands and improve air quality while enhancing the aesthetics of recreational facilities.
- *Small-scale commercial horticulture*: Small-scale commercial horticulture in urban and peri-urban areas refers to cultivating vegetables and fruits primarily for market sale. This type of urban agriculture is prevalent worldwide due to the

high demand for fresh produce. Its proximity to urban markets provides a competitive advantage over rural horticulture. Growers benefit from better infrastructure, technical advice, market information, and potential financial support, focusing primarily on income generation.

- *Street landscaping*: Street landscaping involves beautifying streets with community gardens maintained by residents. These gardens enhance the appearance of streets, purify the air, and create a cleaner environment. Additionally, their strategic placement along streets helps reduce urban stormwater runoff.
- *Therapeutic gardens*: Therapeutic gardens leverage the healing effects of gardening and agriculture to support various treatments. Typically situated within cities at healthcare institutions, these gardens aid in treating mental disorders, autism, Alzheimer's disease, cerebral palsy, and addictions to drugs and alcohol. Contemplative therapeutic gardens, which focus on providing a serene environment, are more common than those involving active gardening practices.
- *Urban aquaculture/aquaponics*: Urban aquaculture involves farming aquatic organisms like fish and plants in urban settings. Aquaponics, a type of urban aquaculture, combines fish farming with hydroponic plant cultivation. In this system, fish waste provides nutrients for plants, while plants filter the water for the fish. Microbes and composting worms convert fish waste into plant food, creating a sustainable cycle.
- *Urban farming*: Urban farming includes various city activities driven by the need to diversify and meet urban demands for recreation and tourism. Urban farms now offer services like landscape management, environmental initiatives, land rental, and direct marketing. They fall into two main categories: farms providing on-site services like leisure, educational, therapeutic, and social farms and local food and environmental farms, which support urban metabolism and the environment.
- *Vertical farming*: Vertical farming refers to growing plants upward using structures such as walls or within buildings. This method can involve soilless systems like hydroponics or traditional soil-based methods. Vertical gardens maximize space for plant production by utilizing vertical surfaces. Practiced by a range of individuals, from urban residents to commercial growers, vertical farming aims to produce fresh food in urban environments efficiently.
- *Zero acreage farms*: High- or low-tech farms focus on space efficiency and resource use and integrate into urban buildings. Examples include vertical farms, rooftop beds, and cellar systems. They primarily grow herbs, greens, mushrooms, insects, and fish. Produce is sold directly to consumers and businesses. Managed by farmers, entrepreneurs, or NGOs, some also feature leisure activities like on-site restaurants [166–170].

7.2.1 Benefits of urban agriculture

Urban farming provides many benefits that enhance food security, promote sustainability, and foster community development in urban areas. These benefits include:

- Increasing food security: Urban areas often form food deserts, where buying good quality or affordable fresh food is difficult. Urban farms can provide fresh produce to low-income individuals who need it most, addressing critical food security issues.
- Creating fresher, healthier foods: Urban farming produces fresh produce closer to where it is consumed, reducing food miles and associated carbon emissions. Produce from urban farms is more likely to be perfectly ripe, nutritious, and in season, contributing to better public health by offering nutrient-dense, healthy alternatives and reducing diet-related health problems.
- Urban regeneration and use of underutilized spaces: Urban farming revitalizes unused or undesirable land, transforming it into green spaces that enhance esthetic appeal and create relaxing community areas. These green spaces can also increase property values, with studies showing surrounding property values rising by nearly 10% within 5 years of establishing community gardens.
- Community involvement: Urban farming brings residents together to work toward a common goal, fostering a sense of belonging and community spirit. Urban farms can offer educational opportunities through tours and workshops, teaching gardening techniques and increasing awareness of food sources. They can also integrate with local businesses, providing fresh produce to restaurants and cafes.
- Efficient land use: Urban farming utilizes techniques like hydroponics, vertical gardens, and rooftop gardens to maximize the use of underutilized urban spaces, providing innovative and efficient solutions to the challenges of growing in the city.
- Economic growth and job creation: As urban farms expand, they can employ more people, offering valuable skills and income opportunities, particularly for low-income individuals. Urban farms stimulate local economies by circulating income within the community, supporting economic growth, and reducing hunger and poverty.
- Less food waste: Urban farms reduce food waste by allowing people to harvest only what they need for immediate consumption, minimizing store and consumer food waste. This fosters a better connection between people and their food sources.
- Water conservation: Urban farming conserves water through efficient irrigation systems, such as timed irrigation and hydroponics, which use significantly less water than conventional farming. Urban farms also reduce water runoff and can collect rainwater from nearby buildings for crop irrigation, further enhancing water conservation.

- Lower investment required: Urban farms require less space and have lower initial setup costs than traditional farms, making urban farming more accessible and feasible for many aspiring farmers [171, 172].

7.3 Urban agriculture and circular economy

Urban agriculture represents a transformative approach within the circular economy, addressing key environmental, social, and economic challenges in urban settings. Traditionally, cities have relied on a linear flow of resources, where food and goods are produced in rural areas, transported to urban centers, consumed, and disposed of, resulting in significant waste and resource depletion. Urban agriculture disrupts this linear model by integrating food production into the urban fabric, creating sustainable and resilient urban ecosystems.

Urban agriculture contributes to the circular economy by redirecting material flows within cities. It transforms waste into valuable resources, positioning itself between catabolic (decomposition) and anabolic (construction) processes. For example, organic waste from households and businesses, typically transported out of the city, can be composted to create nutrient-rich soils for plant growth, effectively closing the material use loop.

In addition, urban agriculture leverages excess urban resources like nutrients, water, and energy. Practices such as rainwater harvesting, wastewater recycling, and solar energy use reduce the need for external inputs, making urban food systems more self-sufficient. Efficient use of these resources minimizes waste and promotes recycling, enhancing city sustainability. Closed-loop water systems, for instance, are integral to urban agriculture. These systems include innovative solutions like hydroponics and aquaponics, which reduce water usage by up to 90% compared to traditional farming methods. Rainwater harvesting and wastewater recycling further conserve water resources, aligning with the principles of a circular economy by ensuring that water is reused and recycled within the urban system.

A significant benefit of urban agriculture within the circular economy is the reduction of transportation-related environmental impacts. Producing food closer to consumers decreases the need for long-distance transportation, reducing greenhouse gas emissions and other pollutants. Shorter supply chains also enhance food freshness and nutritional value.

Waste-to-resource initiatives are pivotal in urban agriculture, transforming organic waste into valuable resources such as compost and bioenergy. These initiatives address the significant environmental challenge posed by food waste. By implementing effective waste management practices, urban agriculture can close the loop on waste, reduce greenhouse gas emissions, and enhance overall sustainability. Composting turns organic waste into nutrient-rich soil, enhancing soil quality and reducing the need for chemical fertilizers. Bioenergy production from organic waste generates renewable energy, contributing to energy sustainability and reducing reliance on fossil fuels.

Socially, urban agriculture fosters community engagement and education, offering residents opportunities to participate in food production and learn about sustainable practices. Community gardens, rooftop farms, and vertical farming systems can become social hubs, promoting intercultural communication and cooperation. Additionally, urban agriculture provides job opportunities, particularly in underserved areas, and enhances food security by ensuring a local supply of fresh produce.

Economically, urban agriculture boosts urban resilience by reducing dependency on global supply chains, which economic crises or natural disasters can disrupt. Producing food within the city reduces waste disposal and transportation costs, while locally grown produce can generate revenue. Furthermore, urban agriculture, as a nature-based solution, enhances urban biodiversity, reduces the urban heat island effect and improves air quality. Integrating urban agriculture into city planning through green roofs, green walls, and urban forests can provide numerous ecosystem services, such as stormwater management, carbon sequestration, and habitat for urban wildlife.

By combining circular economy principles, closed-loop water systems, waste-to-resource initiatives, and nature-based solutions, urban agriculture offers a holistic approach to creating sustainable and resilient urban ecosystems. This integrated approach addresses immediate environmental, social, and economic challenges and promotes long-term sustainability and resilience in urban settings [4, 10, 166, 173].

7.4 Closed-loop water systems in urban agriculture

Urban farming faces a critical challenge in water-scarce regions, where traditional agricultural methods require large quantities of water. This high demand often strains limited resources, worsening water scarcity and declining water quality. To address these issues, urban farmers are increasingly turning to innovative, water-efficient practices that align with closed-loop system principles, promoting more sustainable and resilient water usage.

7.4.1 Hydroponic and aeroponic systems

One of the most significant advancements in water-efficient urban agriculture is the development of hydroponic and aeroponic systems. These systems reduce water usage by up to 90% compared to traditional soil-based farming methods. Hydroponics involves growing plants in a nutrient-rich water solution rather than soil. This method conserves water and allows precise control over nutrient delivery, resulting in faster growth rates and higher yields. Aeroponics takes this further by suspending plants in the air and misting their roots with a nutrient solution, which minimizes water use and enhances oxygen availability to the roots.

The efficiency of these systems lies in their closed-loop nature. Water and nutrients are recirculated, reducing the need for freshwater inputs, and minimizing waste. Additionally, these systems often operate indoors or in controlled environments, protecting plants from pests and diseases and reducing or eliminating the need for chemical pesticides and herbicides. This approach conserves water and promotes healthier, more sustainable food production practices [174, 175].

7.4.2 Rainwater harvesting systems

Rainwater harvesting is another critical component of closed-loop water systems in urban agriculture. These systems capture and store rainwater from rooftops and other surfaces for later use in irrigation. Rainwater harvesting reduces the demand for treated water by decreasing reliance on municipal water supplies. It helps manage stormwater runoff, which can otherwise overwhelm urban drainage systems and lead to flooding and water pollution.

Rainwater harvesting systems can be as simple as rain barrels connected to downspouts or as complex as underground cisterns with filtration and pump systems. The collected rainwater can be used directly for irrigation or processed through filtration systems to ensure its suitability for various agricultural uses. This approach conserves water and aligns with sustainable urban planning by reducing the impact of urbanization on natural water cycles [13–15].

7.4.3 Wastewater recycling

Wastewater recycling represents a significant opportunity for enhancing water sustainability in urban agriculture. Treated wastewater, often reclaimed, or recycled, can be reused for irrigation, reducing the demand for freshwater resources. This practice involves collecting gray water (relatively clean wastewater from baths, sinks, washing machines, and other kitchen appliances) and blackwater (wastewater from toilets), treating it to remove contaminants, and repurposing it for agricultural use.

Advanced treatment technologies, such as membrane bioreactors, ultraviolet disinfection, and reverse osmosis, ensure that the recycled water meets the necessary quality standards for irrigation. Implementing wastewater recycling systems in urban agriculture conserves freshwater and helps manage urban wastewater, reducing the burden on municipal treatment facilities and lowering the risk of water pollution [13–15].

7.4.4 Green roofs and walls

Green roofs and walls are innovative solutions that integrate vegetation into building designs, contributing to water efficiency and urban sustainability. These systems involve growing plants on rooftops and vertical surfaces, which can significantly reduce the urban heat island effect, improve air quality, and enhance stormwater management.

Green roofs are designed with layers that support plant growth including a waterproof membrane, drainage layer, and a growing medium. They absorb and retain rainwater, which is then used by the plants or slowly released, reducing the volume and speed of runoff entering urban drainage systems. This helps conserve water, mitigates flooding risks, and reduces the load on sewage systems during heavy rain events.

Green walls, or vertical gardens, function similarly by incorporating vegetation into the vertical surfaces of buildings. These walls can be designed with integrated irrigation systems that use harvested rainwater or recycled gray water, ensuring efficient water use. The vegetation on green walls helps insulate buildings, reducing energy consumption for heating and cooling and improves air quality by capturing pollutants and producing oxygen [176].

7.4.5 Aquaponics systems

Aquaponics combines aquaculture (raising fish) and hydroponics in a closed-loop system where fish waste provides nutrients for plants, and the plants help filter and purify the water for the fish. This symbiotic relationship allows for efficient resource use and reduces the need for chemical fertilizers.

In an aquaponics system, water from the fish tanks is circulated through plant beds, where microorganisms convert fish waste into nutrients that plants can absorb. The cleaned water is then recirculated back to the fish tanks. This closed-loop system conserves water; the only significant loss is through evaporation and plant uptake. Additionally, aquaponics systems can produce fish and vegetables, enhancing food security and sustainability in urban environments [13–15].

7.4.6 Subsurface irrigation

Subsurface irrigation systems deliver water directly to the root zone of plants through buried tubing or emitters. This method reduces water loss due to evaporation and runoff, ensuring that more water reaches the plant roots where needed most. Subsurface irrigation can be particularly effective in urban agriculture settings where space is limited and efficient water use is critical.

By delivering water directly to the roots, subsurface irrigation systems minimize weed growth and reduce the risk of soil-borne diseases. These systems can be integrated with soil moisture sensors and automated controllers to optimize watering schedules, further enhancing water efficiency and crop yields [14, 15].

7.4.7 Drip irrigation

Drip irrigation is a widely used method in urban agriculture due to its efficiency and effectiveness. This system delivers water directly to the base of each plant through a network of tubes and emitters, minimizing water waste and ensuring precise delivery to the root zone. Drip irrigation systems can be customized to suit various urban farming setups from small garden plots to large rooftop farms.

The benefits of drip irrigation include reduced water usage, lower weed growth, and improved plant health. By providing a consistent and controlled water supply, drip irrigation helps maintain optimal soil moisture levels, promoting healthy root development and reducing the risk of water stress [14, 15].

7.4.8 Water-saving technologies and practices

Urban farmers are increasingly adopting various water-saving technologies and practices to enhance the sustainability of their operations. These include
- Soil moisture sensors: These devices measure soil moisture levels and provide real-time data to optimize irrigation schedules, ensuring that plants receive the right amount of water without waste.
- Mulching: Applying a layer of organic or inorganic material on the soil surface helps retain soil moisture, reduce evaporation, and suppress weed growth.
- Rain gardens: Designed to capture and absorb rainwater runoff, rain gardens are shallow, vegetated basins that allow water to infiltrate the soil, reducing runoff and recharging groundwater supplies.
- Permeable pavements: These surfaces allow water to pass through, reducing runoff and promoting groundwater recharge. They can be integrated into urban farms to enhance water management [14, 15, 176].

7.4.9 Future trends in water management for urban agriculture

As urban agriculture continues to evolve, new trends and technologies are emerging to enhance water management practices further. These include
- Smart irrigation systems: Leveraging the Internet of things and data analytics, smart irrigation systems use weather forecasts, soil moisture data, and plant

water requirements to automate and optimize irrigation, maximizing efficiency and reducing waste.
– Advanced water recycling: Emerging technologies for water recycling, such as decentralized wastewater treatment systems and on-site gray water treatment units, offer scalable solutions for urban farms to recycle and reuse water efficiently.
– Desalination technologies: In coastal cities, desalination technologies can provide an additional source of irrigation water. Advanced, energy-efficient desalination systems are becoming more viable for integration into urban agricultural practices.
– Vertical and indoor farming innovations: Innovations in vertical and indoor farming, such as advanced LED lighting, climate control systems, and automated nutrient delivery, are making these methods more water-efficient and sustainable [137].

Case 7.1 Reflo's impact on sustainable water use and urban agriculture in Milwaukee

Reflo, a 501(c)(3) nonprofit based in Milwaukee, Wisconsin, aims to catalyze sustainable water use, green infrastructure, and equitable water resource management to support triple-bottom-line outcomes grounded in strong partner relationships. Each year, Reflo supports community green infrastructure projects that harvest stormwater for urban agriculture and other uses, resulting in the construction of large underground cisterns for Milwaukee's urban farming scene. These projects provide environmental, economic, social, and artistic benefits. Reflo pioneered new codes and ordinances to allow freestanding rainwater harvesting structures in Milwaukee. In 2018, over 150 volunteers built a 20,000-gallon cistern at Alice's Garden in 2 h, which collects water from the Brown Street Academy playground via a bioswale and uses a solar-powered pump to irrigate community plots. Reflo collaborated with partners to design and construct a 40,000-gallon underground cistern at Cream City Farms, where stormwater is filtered in bioswales and pumped via a solar pump to irrigate vegetable crops. Additionally, Reflo supported an award-winning project to create a pavilion that collects rainwater for raised bed gardens at Milwaukee's Guest House men's homeless shelter [177].

7.5 Waste-to-resource initiatives

Waste-to-resource initiatives are essential in urban agriculture, transforming organic waste into valuable resources such as compost and bioenergy. These initiatives address the significant environmental challenge posed by food waste, which the United Nations Environment Programme estimates to be 931 million tons annually. By implementing effective waste management practices, urban agriculture can close the loop on waste, reduce greenhouse gas emissions, and enhance overall sustainability [178].

7.5.1 Composting

Composting is a cornerstone of waste-to-resource initiatives in urban agriculture. This process involves the aerobic decomposition of organic waste materials by microorganisms including food scraps, yard waste, and paper products. The resulting product, compost, is a nutrient-rich soil amendment that enhances soil quality and fertility.

7.5.1.1 Benefits of composting

Composting offers numerous advantages for urban agriculture, ranging from enhancing soil health to reducing greenhouse gas emissions and supporting sustainable waste management practices:
- Soil health: Compost improves soil structure, aeration, and water retention, which are critical for plant growth. It adds organic matter to the soil, promoting the activity of beneficial soil organisms.
- Nutrient recycling: Composting recycles nutrients back into the soil, reducing the need for chemical fertilizers. This contributes to more sustainable farming practices and reduces agricultural runoff, which can pollute water bodies.
- Waste reduction: Composting reduces methane emissions, a potent greenhouse gas generated by anaerobic decomposition in landfills, by diverting organic waste from landfills. This helps mitigate climate change.
- Economic benefits: Composting can reduce waste disposal costs for cities and provide a valuable product that can be sold or used within urban farms, creating economic opportunities [179].

7.5.1.2 Types of composting systems

Various composting systems are available, each suited to different scales and contexts, from individual households to large commercial operations:
- Backyard composting: Suitable for individual households, backyard composting involves creating compost piles or using bins to manage organic waste. It is a simple and effective way to recycle kitchen scraps and yard waste.
- Community composting: This involves neighborhood or community-based composting programs in which residents collectively manage organic waste. These programs can foster community engagement and environmental awareness.
- Commercial composting: Larger-scale operations are managed by municipalities or private companies. These facilities handle significant volumes of waste and can process a wider variety of materials including food waste from restaurants, supermarkets, and food processing facilities [179].

7.5.1.3 Best practices for composting

Implementing best practices in composting ensures efficient decomposition, maximizes nutrient recycling, and minimizes potential issues such as odors and pests:

- Balancing materials: Effective composting requires a balance of green (nitrogen-rich) and brown (carbon-rich) materials. Green materials include fruit and vegetable scraps, coffee grounds, and grass clippings, while brown materials include leaves, straw, and cardboard.
- Maintaining moisture: Compost piles should be kept moist but not waterlogged. Proper moisture levels are essential for microbial activity.
- Aeration: Regularly turning the compost pile ensures adequate oxygen supply, promoting aerobic decomposition and preventing odors.
- Temperature monitoring: Compost piles should reach temperatures between 131 °F and 170 °F to kill pathogens and weed seeds. Monitoring and managing the temperature is crucial for effective composting [179, 180].

7.5.2 Bioenergy production

Bioenergy production is another critical component of waste-to-resource initiatives in urban agriculture. This process converts organic waste into renewable energy sources such as biogas and biofuel, providing a sustainable energy alternative and reducing dependence on fossil fuels.

7.5.2.1 Biogas production

Biogas is produced through anaerobic digestion, where microorganisms break down organic matter without oxygen. This process generates biogas, primarily composed of methane and carbon dioxide, and a nutrient-rich digestate that can be used as a fertilizer.

7.5.2.2 Benefits of biogas production

Biogas production provides significant advantages including generating renewable energy, reducing waste volumes, recycling nutrients, and offering economic opportunities:

- Renewable energy: Biogas can be used for heating, electricity generation, and as a vehicle fuel, providing a renewable energy source that reduces greenhouse gas emissions.
- Waste management: Anaerobic digestion reduces the volume of organic waste, decreasing landfill use and associated environmental impacts.
- Nutrient recycling: The digestate produced during anaerobic digestion is a valuable fertilizer, returning nutrients to the soil and reducing the need for synthetic fertilizers.

– Economic opportunities: Biogas production can create jobs and generate revenue by selling biogas and digestate [4].

7.5.3 Types of biogas systems

Biogas systems come in various scales and configurations, from small household digesters to large industrial facilities, each designed to meet different waste processing and energy production needs:
– Small-scale digesters: Suitable for individual households or small farms, these systems can process kitchen waste, animal manure, and crop residues to produce biogas for local use.
– Community-scale digesters: Larger systems that serve neighborhoods or communities, processing organic waste from multiple sources and providing biogas for local energy needs.
– Industrial-scale digesters: Large facilities that handle significant volumes of organic waste from municipal sources, food processing plants, and agricultural operations. These systems produce substantial biogas and can contribute to regional energy grids [181, 182].

7.5.3.1 Best practices for biogas production
Adhering to best practices in biogas production ensures optimal feedstock management, temperature control, retention time, and system maintenance, maximizing biogas yield and efficiency:
– Feedstock management: Selecting appropriate feedstock materials and maintaining a consistent supply is crucial for efficient biogas production.
– Temperature control: Anaerobic digesters operate best at mesophilic (95–105 °F) or thermophilic (125–140 °F) temperatures. Maintaining optimal temperatures ensures efficient microbial activity.
– Retention time: Ensuring organic matter remains in the digester long enough to be fully decomposed maximizes biogas yield and quality.
– System maintenance: Regular maintenance of digesters, including cleaning and monitoring gas production, is essential for reliable operation [183–185].

7.5.4 Integration of waste-to-resource initiatives in urban agriculture

Integrating waste-to-resource initiatives into urban agriculture requires careful planning and coordination among various stakeholders including city planners, farmers, waste management professionals, and community members. Effective integration of waste-to-resource initiatives in urban agriculture requires robust policy support, infrastructure development, education and outreach, stakeholder collaboration, and the adoption of innovative technologies:

- Policy support: Municipal policies that promote composting and bioenergy production, such as incentives for waste diversion and support for renewable energy projects, are essential for successful implementation.
- Infrastructure development: Developing the necessary infrastructure, such as composting facilities and biogas plants, is crucial for processing organic waste efficiently.
- Education and outreach: Educating residents and businesses about the benefits of composting and bioenergy production can increase participation and support for waste-to-resource initiatives.
- Collaboration: Collaboration among various stakeholders, including local governments, nonprofit organizations, and private sector partners, can enhance the effectiveness of waste-to-resource initiatives.
- Innovation and technology: Embracing innovative technologies and practices, such as smart composting systems and advanced biogas digesters, can improve the efficiency and sustainability of waste-to-resource initiatives [4].

Case 7.2 Stockholm's Food Waste Initiative transforming waste into biogas

As of January 1, 2023, Stockholm mandated that all households, offices, and businesses separate food waste from other waste, aiming to transform it into biogas. This initiative positioned Stockholm ahead of and behind current practices in waste management. While many Swedish municipalities have adopted food waste collection since the 1990s, Stockholm lagged in comprehensive implementation. With 40% of waste being food-related, the potential for biogas production was significant. Stockholm Vatten och Avfall highlighted that five kilos of food waste could power a car for almost a mile, and food waste from 3,000 people could run a city bus for a year. Despite these benefits, only 30% of the city's food waste was sorted for biogas production. Stockholm sought to enhance resource utilization, reduce greenhouse gas emissions, and promote sustainable energy by making sorting compulsory. Property owners were responsible for ensuring compliance, with some exceptions until July 2024, emphasizing the city's commitment to environmental stewardship [186].

7.6 Urban agriculture as a nature-based solution

Urban agriculture is increasingly recognized as a vital component of sustainable urban development. When integrated into the urban system and tailored to meet the needs of urban residents, it offers numerous benefits beyond food production. Urban agriculture provides various ecosystem services, grouped into provisioning, regulating, cultural, and habitat services. These contributions make urban environments more resilient, sustainable, and livable.

7.6.1 Provisioning services

One of the primary benefits of urban agriculture is its ability to provide provisioning services. These services include food, fiber, and biomass production, essential resources for urban populations. By growing food locally, urban agriculture reduces the dependency on food transported from rural areas, often involving long distances and significant carbon emissions. This local food production enhances food security by ensuring that urban residents can access fresh, nutritious, and affordable produce. This is particularly important in areas classified as food deserts, where access to fresh food is limited.

In addition to food, urban agriculture can produce fiber and biomass. These materials can be used for various purposes, including textile production and bioenergy, as raw materials for other industries. By diversifying the types of products generated, urban agriculture supports a more robust and resilient local economy capable of adapting to changes in demand and market conditions. Furthermore, urban farms and gardens enhance pollination services by cultivating diverse plant species that attract pollinators like bees and butterflies. These pollinators are crucial for the reproduction of many plants, including those grown in urban farms, thus supporting the overall biodiversity and ecological health of urban areas [36].

7.6.2 Regulating services

Urban agriculture provides several critical regulating services that help maintain ecological balance and promote sustainability in urban areas. These services include climate regulation, water management, and waste recycling.

7.6.2.1 Climate regulation
Urban agriculture contributes to climate regulation by mitigating the urban heat island effect. Plants and green spaces absorb sunlight and release moisture through transpiration, which cools the air and reduces temperatures in urban areas. This cooling effect is significant in densely built cities, where heat accumulates and creates uncomfortable living conditions. Additionally, urban agriculture helps sequester carbon by absorbing carbon dioxide through photosynthesis and storing it in plant biomass and soils. This process reduces greenhouse gas concentrations in the atmosphere, helping mitigate climate change [36].

7.6.2.2 Water management

Urban agriculture plays a significant role in urban water management. Agricultural soils support groundwater recharge by allowing rainwater to infiltrate the ground, which helps replenish groundwater supplies and reduce the risk of water shortages. Urban farms and gardens also aid in stormwater retention by capturing and absorbing rainwater, reducing runoff, and preventing flooding. Green roofs, rain gardens, and permeable pavements associated with urban agriculture enhance these benefits by providing additional water infiltration and storage surfaces [14, 15].

7.6.2.3 Waste recycling

Urban agriculture supports waste recycling through composting and using organic waste as fertilizers. Composting transforms organic waste, including food scraps and yard waste, into nutrient-rich compost that enhances soil fertility and structure. This process reduces the need for chemical fertilizers, decreases landfill waste, and lowers greenhouse gas emissions from waste decomposition. Additionally, urban farms can utilize treated wastewater for irrigation, further conserving water resources and reducing the demand for fresh water [15].

7.6.3 Cultural services

Urban agriculture offers numerous cultural services that enhance urban communities' social and cultural fabric. These services include recreational opportunities, educational programs, and the promotion of community engagement and social cohesion.

7.6.3.1 Recreational opportunities

Urban farms and gardens provide green spaces for recreation and relaxation. These areas offer residents opportunities for physical activity, social interaction, and connection with nature, which can improve overall well-being and quality of life. Recreational activities such as gardening, walking, and community events in urban agricultural spaces foster a sense of community and belonging among residents [36].

7.6.3.2 Educational programs

Urban agriculture serves as a valuable educational resource and natural laboratory. Schools, universities, and community organizations can use urban farms and gardens to educate people about agriculture, sustainability, and environmental stewardship. These educational activities promote awareness and understanding of sustainable food systems and the importance of local food production. In addition to formal education, urban agriculture provides opportunities for skills development in areas such as gardening, farming, food processing, and entrepreneurship. These skills can be particularly valuable for young people and those seeking new career opportunities [36].

7.6.3.3 Community engagement and social cohesion
Urban agriculture fosters community engagement and social cohesion by bringing people together to work toward common goals. Community gardens and urban farms serve as social hubs where residents can collaborate, share knowledge, and support each other. These spaces promote intercultural communication and cooperation, helping to build strong, inclusive communities. Urban agriculture also provides volunteerism and civic participation opportunities, encouraging residents to take an active role in improving their neighborhoods [36].

7.6.4 Habitat services

Urban agriculture contributes to habitat services by creating and maintaining habitats for various species, supporting biodiversity, and enhancing ecological resilience in urban areas.

7.6.4.1 Habitat creation
Urban farms and gardens create habitats for various species including plants, insects, birds, and small mammals. Green roofs, walls, and other agricultural installations provide homes for urban wildlife, supporting biodiversity and ecological health. By incorporating native plants and creating diverse habitats, urban agriculture helps preserve and enhance urban biodiversity [36].

7.6.4.2 Biodiversity conservation
Urban agriculture promotes biodiversity conservation by cultivating various plant species including heirloom and indigenous crops. This diversity is crucial for resilient food systems and ecological health. By preserving and promoting a wide range of plant species, urban agriculture helps protect genetic resources and supports conservation efforts. Additionally, urban farms can serve as refuges for pollinators and other beneficial insects, which are essential for healthy ecosystems [36].

7.6.5 Integrating urban agriculture into green infrastructure

Integrating these practices into urban planning and development strategies is essential to maximize the benefits of urban agriculture as a nature-based solution. This integration involves creating supportive policies, investing in infrastructure, and fostering collaboration among various stakeholders:

- Supportive policies: Municipal policies that promote urban agriculture and nature-based solutions are crucial for successful integration. These policies can include zoning laws that allow for urban farming, incentives for green roofs and rain gardens, and support for community gardening programs. By creating a favorable policy environment, cities can encourage the adoption of urban agriculture and nature-based solutions.
- Infrastructure investment: Infrastructure is essential for supporting urban agriculture and nature-based solutions. This includes developing facilities for composting, rainwater harvesting, and wastewater treatment as well as creating green spaces such as parks, community gardens, and urban farms. Infrastructure investment can enhance urban areas' capacity to implement and benefit from nature-based solutions.
- Stakeholder collaboration: Collaboration among various stakeholders, including city planners, community organizations, businesses, and residents, is critical for successfully integrating urban agriculture and nature-based solutions. By working together, stakeholders can share resources, knowledge, and expertise and coordinate efforts to create sustainable and resilient urban environments.
- Public awareness and education: Raising public awareness and educating the public about the benefits of urban agriculture and nature-based solutions is essential for gaining community support and participation. Educational programs, workshops, and outreach activities can help residents understand the value of these practices and how they can contribute to urban sustainability.
- Monitoring and evaluation: Implementing systems for monitoring and evaluating the effectiveness of urban agriculture and nature-based solutions projects is essential for continuous improvement. By assessing the impacts of these initiatives on environmental, social, and economic outcomes, cities can identify best practices, address challenges, and ensure that they achieve their sustainability goals [36, 166, 176].

Case 7.3 Benefits of food forests and urban farms in San Antonio

In collaboration with the Food Policy Council of San Antonio and three city departments (Innovation, Metro Health, and Sustainability), Stanford University's Natural Capital Project released the study "Vibrant Land: The Benefits of Food Forests and Urban Farms in San Antonio." The study quantifies urban agriculture's benefits across San Antonio, using existing sites to inform farming practices and crop selection. The study guides investment decisions to benefit vulnerable communities by linking food-insecure households with underutilized public land. Urban agriculture in these areas could address food insecurity, urban heat, and flooding. The study highlights cobenefits such as urban cooling, carbon sequestration, green space access, floodwater retention, and nutrient retention provided by food forests and urban farms. Funded by NASA's Environmental Equity and Justice program, the project develops a toolkit for urban planners to explore and ensure equitable distribution of these benefits [187].

7.7 Strategies to promote urban agriculture

Cities can employ numerous approaches to foster urban agriculture, such as establishing favorable policies, providing incentives for urban farming, and delivering programs, financial support, and public land to aid urban agriculture initiatives, as detailed in Table 7.1 [188–193].

Table 7.1: Strategies to promote urban agriculture.

Strategy	Description
Policy development	Zoning laws play a pivotal role in supporting urban agriculture. These policies convert underutilized spaces into productive agricultural areas, promoting local food production and green space enhancement by allowing community gardens and urban farms in various zones. Effective policy development can transform urban landscapes and support sustainable food systems.
Financial incentives	Financial incentives are crucial for encouraging urban farming. Tax breaks, grants, and low-interest loans reduce financial burdens, making it easier for individuals and organizations to start and sustain urban agriculture projects. These incentives boost local food production, stimulate economic growth, and encourage investment in sustainable practices.
Public land allocation	Allocating public land for urban agriculture supports the development of community gardens and urban farms. This approach maximizes the use of available land, promotes local food production, and enhances community engagement. Public land allocation ensures that urban agriculture projects have the space needed to thrive and benefit the community.
Educational programs	Educational programs are essential for urban agriculture. Residents can learn effective farming techniques, sustainability practices, and nutritional benefits by offering workshops, training sessions, and resources. These programs empower communities to engage in and support urban agriculture, fostering a culture of learning and growth around local food production.
Infrastructure support	Investing in infrastructure like irrigation systems, greenhouses, and composting facilities significantly enhances urban agriculture. These investments ensure efficient resource use, increase productivity, and support sustainable farming practices within urban settings. Infrastructure support is critical to creating resilient and productive urban agriculture systems.

Table 7.1 (continued)

Strategy	Description
Community partnerships	Building community partnerships with local organizations, schools, and businesses strengthens urban agriculture initiatives. These collaborations provide resources, knowledge, and support, fostering a united effort toward sustainable urban farming and community well-being. Community partnerships create a network of support that enhances the impact of urban agriculture.
Market access	Facilitating market access for urban farmers is vital. Providing entry to farmers' markets, food cooperatives, and farm-to-table programs ensures that urban agriculture products reach consumers. This support helps build local economies, promote sustainable food systems, and ensure that urban farmers have viable outlets for their produce.
Technical assistance	Offering technical assistance helps urban farmers address challenges. Consulting services guide regulatory compliance, pest management, and crop selection, ensuring successful and sustainable urban farming operations. Technical assistance is crucial for helping urban farmers navigate the complexities of modern agriculture.
Research and development	Funding research and development for innovative urban agriculture practices like vertical farming, hydroponics, and aquaponics drives progress. These advancements improve efficiency, productivity, and sustainability in urban farming. Research and development efforts are essential for pioneering new methods to scale up urban agriculture.
Advocacy and awareness campaigns	Advocacy and awareness campaigns are essential for promoting urban agriculture. These initiatives highlight its benefits, encourage community involvement, and drive policy changes at various government levels. Effective campaigns foster a supportive environment for urban farming, raising public awareness and gaining support for sustainable urban food systems.

Case 7.4 Chesapeake Bay Trust and District of Columbia Department of Energy and Environment DC Urban Agriculture Small Award Program

The Chesapeake Bay Trust, in collaboration with the District of Columbia Department of Energy and Environment Office of Urban Agriculture, administers the DC Urban Agriculture Small Award Program. This initiative aims to boost urban agriculture by providing grants of up to $10,000 to support the infrastructure and operations of urban farms, particularly benefiting socially disadvantaged farmers and communities with limited access to fresh food. Eligible applicants include nonprofit organizations, small businesses, and educational institutions within the District. The funding supports projects that address at least two key goals: increasing food crop production, processing, and distribution; improving access to fresh foods; constructing facilities for agricultural education; accelerating the business and production capacity for socially disadvantaged farmers; and advancing sustainable agricultural efforts. Urban farms receive crucial support through this program to expand operations, increase food access, and promote sustainable agricultural practices [194].

Chapter 8
Financing the circular economy investment strategies and partnerships in the water-food nexus

Abstract: This chapter examines the role of financing in promoting the circular economy within the water-food nexus. It covers public financing mechanisms including government grants, international funding programs, and private options such as venture capital (VC), private equity (PE), and green bonds. It also discusses innovative financial instruments like blended finance and crowdfunding. Emphasizing the importance of strategic partnerships, it includes case studies of successful collaborations. The chapter concludes by identifying future research needs and offering recommendations for stakeholders, encouraging policymakers, investors, businesses, and civil society to support sustainable investments and practices for a resilient and sustainable future.

8.1 Introduction

This chapter examines the role of financial investments and strategic partnerships in advancing the circular economy within the water-food nexus. It explores various financing mechanisms and instruments, highlighting how private and public funding can support the development and scaling of circular practices and technologies. The chapter focuses on identifying effective strategies for mobilizing capital and fostering collaborations that contribute to the sustainable management of water and food resources. It illustrates the potential for scaling up successful models and overcoming financial barriers through case studies, examples, and innovative financing approaches. Additionally, it emphasizes the importance of strategic partnerships in leveraging resources and expertise to achieve sustainable outcomes.

8.2 The role of finance in the circular economy

Finance is critical in advancing the circular economy, particularly within the water-food nexus. It provides the essential resources to develop, implement, and scale sustainable practices and technologies. Finance supports innovation by facilitating the transition from traditional linear models to more circular and resource-efficient systems. It helps address the challenges associated with resource management, environmental sustainability, and economic resilience. The effective allocation and manage-

https://doi.org/10.1515/9783111341385-008

ment of financial resources are crucial to realizing the full potential of circular economy principles within the water-food nexus.

8.2.1 Definition and significance of financing in the circular economy

Financing in the circular economy refers to providing funds and capital to support initiatives, projects, and businesses that adhere to reducing waste, reusing resources, and recycling materials. This financing is crucial for developing and scaling up sustainable practices and technologies that minimize environmental impact and promote resource efficiency. Transitioning from a traditional linear economy to a circular economy requires significant investment to develop new technologies, infrastructure, and business models to sustain this shift [10].

8.2.1.1 Importance of capital for innovation and scalability in the water-food nexus

Capital is essential for driving innovation and scalability within the water-food nexus. Investments are needed to develop and implement advanced technologies and practices that improve water efficiency and sustainable food production. Innovations such as precision agriculture, efficient irrigation systems, wastewater recycling, and sustainable aquaculture require substantial funding. Moreover, scaling up these innovations to achieve widespread adoption and impact necessitates continuous financial support. Adequate financing ensures that these solutions can be developed, tested, and deployed on a large scale, contributing to the overall sustainability of the water-food nexus [4].

8.2.2 Key stakeholders: private sector, public sector, and international organizations

Several key stakeholders play vital roles in financing the circular economy within the water-food nexus:
- *Private sector*: Banks, venture capital (VC) firms, private equity PE) funds, and corporations. The private sector provides significant funding for research, development, and commercialization of new technologies and practices. Companies often invest in sustainable projects as part of their corporate social responsibility (CSR) initiatives or to gain competitive advantages in emerging markets.
- *Public sector*: Governments and public institutions offer grants, subsidies, and policy support to encourage the development and adoption of circular economy practices. They also facilitate public-private partnerships (PPPs) to leverage private capital for public projects.

International organizations: Entities like the World Bank, the International Monetary Fund (IMF), and various United Nations agencies provide funding and technical assistance for sustainable development projects worldwide. These organizations play a crucial role in mobilizing resources and fostering international collaboration [4, 195, 196].

8.2.2.1 Roles and responsibilities
Each stakeholder has distinct roles and responsibilities:
– *Private sector*: Provides capital, expertise, and innovation, engages in developing and scaling up new technologies and practices, and assumes risks associated with investing in new ventures.
– *Public sector*: Creates enabling environments through policies and regulations, provides financial incentives and supports infrastructure development, and facilitates collaboration between private and public entities.
– *International organizations*: Mobilize global resources, provide financial assistance and technical expertise, promote best practices, and facilitate knowledge exchange.

8.2.2.2 Examples of influential stakeholders in the water-food sector
Several stakeholders have significantly impacted the water-food sector:
– *Bill and Melinda Gates Foundation*: Funds innovative agricultural projects to improve food security and sustainability.
– *International Finance Corporation (IFC)*: Invests in private-sector projects that promote sustainable development.
– *European Investment Bank (EIB)*: This institution funds projects that improve water management and food production in Europe and beyond [197–199].

8.2.3 Current state of financing in the water-food nexus

The current state of financing in the water-food nexus shows progress and challenges. Significant investments have been made in various sustainable projects, yet there is still a considerable need for increased funding and better resource allocation. While there are successful examples of funded projects, such as sustainable irrigation systems and wastewater recycling plants, many initiatives still struggle to secure adequate financing [4, 200, 201].

8.2.3.1 Overview of existing investments
Existing investments in the water-food nexus come from diverse sources including government grants, PE, VC, and international funding programs. These investments

support various projects from technological innovations to large-scale infrastructure developments. However, the distribution of funds is often uneven, with certain regions and sectors receiving more attention than others [4].

8.2.3.2 Gaps and opportunities for improvement
Several gaps and opportunities exist for improving financing in the water-food nexus:
–   *Gaps*: Insufficient funding for small-scale and grassroots projects. Lack of access to capital for developing countries. Inadequate financial instruments tailored to the needs of the water-food sector.
–   *Opportunities*: Increasing awareness and interest in sustainable investments; developing new financial instruments such as green bonds and blended finance; strengthening international cooperation to mobilize resources more effectively [202–205].

8.3 Public financing mechanisms

Public financing mechanisms are essential for promoting sustainable practices within the water-food nexus. These mechanisms, which include government grants, subsidies, and policy incentives, provide crucial financial support for developing and implementing circular economy initiatives. By allocating public funds to projects that enhance resource efficiency, reduce waste, and support sustainable agricultural practices, these financing mechanisms help drive the transition toward a more resilient and sustainable economy. Public financing is particularly significant in enabling large-scale infrastructure projects, research, and development and adopting new technologies that may not be immediately profitable but are vital for long-term sustainability.

8.3.1 Government grants and subsidies

Government grants and subsidies are vital in supporting water-food projects, providing essential funding to promote sustainable practices and innovations. National and regional governments typically offer these financial aids to encourage projects that address critical issues such as water conservation, efficient irrigation, sustainable agriculture, and food security [4].

8.3.1.1 Types of grants and subsidies available for water-food projects
There are several types of grants and subsidies available for water-food projects, each designed to provide financial support and incentives to promote sustainable practices and innovation in this sector:

- *Direct grants*: These are nonrepayable funds designed to support specific projects or initiatives. They often target research and development, pilot projects, and the implementation of new technologies.
- *Subsidized loans*: These are loans offered at reduced interest rates to make financing more affordable for water-food projects. Subsidized loans help lower the cost of capital, making it easier for projects to become financially viable.
- *Tax incentives*: Governments may offer tax credits or deductions for sustainable water and food project investments. These incentives reduce the overall tax burden for businesses and individuals investing in such initiatives.
- *Rebate programs*: These programs provide cash rebates for purchasing and installing water-efficient technologies or sustainable agricultural practices. Rebates help offset the initial costs, making it more attractive for stakeholders to adopt innovative solutions [4].

8.3.2 Public-private partnerships

PPPs are collaborative agreements between government entities and private sector companies designed to finance, build, and operate projects that serve the public interest. PPPs leverage the strengths of both sectors, combining public oversight with private sector efficiency and innovation.

8.3.2.1 Definition and structure of PPPs

A PPP typically involves a long-term contract where a private entity is responsible for designing, financing, constructing, and operating a project. The public sector retains ownership and oversight, ensuring the project's objectives align with public policy goals. The private entity recoups its investment through payments from the public sector or user fees.

8.3.2.2 Benefits of PPPs in the water-food sector

PPPs offer several benefits in the water-food sector, enhancing project viability and sustainability through a combination of private investment, efficiency, innovation, and risk sharing:

- *Access to private capital*: PPPs mobilize private investment, reducing the financial burden on public budgets.
- *Efficiency and innovation*: Private-sector involvement often leads to improved efficiency and adoption of innovative technologies and practices.
- *Risk sharing*: Project development and operation risks are shared between public and private partners, making projects more viable and sustainable [14, 15].

Case 8.1 California's Agriculture Water Use Efficiency Program
The Department of Water Resources (DWR) and the California Department of Food and Agriculture (CDFA) collaboratively launched the Agriculture Water Use Efficiency CDFA-DWR program. This competitive grant initiative aims to showcase the multifaceted benefits of improving water conveyance and on-farm water use efficiency while reducing greenhouse gas emissions. The program addresses critical objectives such as enhancing water use efficiency, conserving water, reducing greenhouse gas emissions, protecting groundwater, and promoting the sustainability of agricultural operations. Eligible applicants for the program include public agencies, utilities, tribes, nonprofits, and investor-owned utilities. An example of the program's impact is the grant awarded on September 13, 2017, to the North San Joaquin Water Conservation District. The $3 million grant was allocated to upgrade the district's water conveyance system, enabling pressurized, on-demand water delivery to growers. In addition, 19 grower coapplicants received $1.65 million in State Water Efficiency and Enhancement Program funds for on-farm efficiency improvements. The benefits of these combined projects are significant. They include annual water savings of 1,000 acre-feet through reduced system losses and 1,800 acre-feet through on-farm efficiency improvements. Greenhouse gas emissions are expected to decrease by 370 metric tons of CO_2 equivalent per year. Furthermore, the projects will enhance groundwater recharge by 12,000 acre-feet annually, improve water and air quality, and boost crop health [206].

8.3.3 International funding programs and development banks

International funding programs and development banks are crucial in supporting sustainable water-food projects, particularly in developing countries. These institutions provide financial resources, technical assistance, and policy advice to help countries implement sustainable practices and infrastructure.

8.3.3.1 Overview of multilateral and bilateral funding initiatives
One of the main challenges in achieving sustainable water-food security is the lack of sufficient and stable funding for projects that address the interlinkages between water and food systems. Two international funding initiatives – multilateral and bilateral – can support the water-food nexus.

8.3.3.1.1 Multilateral funding initiative
Multilateral funding initiatives are a form of international cooperation that aim to address global challenges and promote sustainable development. These initiatives involve multiple countries and international organizations contributing financial re-

sources, technical expertise, and policy guidance to support projects in various regions and sectors. Some benefits of multilateral funding include leveraging diverse sources of finance, enhancing coordination and coherence, fostering knowledge sharing and innovation, and strengthening accountability and transparency. Some examples of multilateral funding initiatives that support the water-food nexus are:

- *The World Bank*: The World Bank is a global development institution that provides loans, grants, and technical assistance to low– and middle-income countries. It supports projects that improve water management, irrigation, agriculture, food security, and climate resilience.
- *The International Monetary Fund (IMF)*: The IMF is an international organization that oversees the global monetary system and provides financial assistance and policy advice to its member countries. The IMF supports macroeconomic stability, fiscal sustainability, and structural reforms to enhance water-food security and reduce poverty.
- *The Green Climate Fund (GCF)*: The GCF is a global fund that supports developing countries in mitigating and adapting to climate change. The GCF finances projects that address the impacts of climate change on water and food systems such as droughts, floods, salinization, and crop losses [207–209].

8.3.3.1.2 Bilateral funding initiatives
These involve financial agreements between two countries, where one country provides funding and technical assistance to support projects in the recipient country.

Some examples of bilateral funding initiatives that support the water-food nexus are:

- The USAID Feed the Future initiative aims to reduce hunger and poverty by investing in agricultural development, nutrition, and resilience. The initiative supports projects that improve water management, irrigation, and watershed protection in various regions such as the Sahel, Central America, and South Asia.
- The Japan International Cooperation Agency Water and Food Security Project in Kenya provides technical assistance and equipment to improve water supply, sanitation, irrigation, and crop production in arid and semiarid areas. The project also promotes community participation, gender equality, and environmental sustainability [210–212].

Case 8.2 Climate Change Adaptation in Rural Areas of India
The Climate Change Adaptation in Rural Areas of India (CCA-RAI) project, commissioned by the German Federal Ministry for Economic Cooperation and Development and executed by the Ministry of Environment and Forests (MoEF), operated from 2008 to 2019. This initiative aimed to integrate climate change adaptation into sector policies and rural development programs across India, targeting the states of Madhya Pradesh, Rajasthan, Tamil Nadu, and West Bengal. India, highly vulnerable to climate

change, faces severe impacts on natural resources, livelihoods, and infrastructure. The Indian Government's 2008 National Action Plan on Climate Change laid the groundwork for addressing these challenges. CCA-RAI sought to support this by integrating climate adaptation into policy and development planning to safeguard rural livelihoods and promote sustainable development. The project conducted state-level vulnerability and risk assessments, developed and demonstrated adaptation measures in nine project sites, and introduced a climate-proofing tool to ensure public sector program sustainability. It collaborated with NGOs, research institutes, and universities to improve socioeconomic conditions and environmental services, enhancing community adaptive capacities. In addition, the project trained policymakers and practitioners, developing a team of skilled trainers and building capacities across various levels. It also assisted in developing climate action plans for 16 states and two union territories, with 11 endorsed by the MoEFCC. CCA-RAI significantly advanced India's preparedness and resilience to climate change through these efforts, demonstrating effective strategies for rural adaptation and sustainable development [213].

8.4 Private financing mechanisms

Private financing mechanisms are essential in advancing the circular economy within the water-food nexus. These mechanisms, including VC, PE, green bonds, sustainability-linked loans, CSR, and impact investing, provide the necessary capital and support for developing and scaling innovative technologies and sustainable practices. This section explores these mechanisms, highlighting their roles, benefits, and examples in the water-food sector.

8.4.1 Role in funding innovative circular technologies in the water-food nexus

VC and PE are pivotal in financing innovative technologies and businesses that contribute to the circular economy within the water-food nexus. These investors provide the capital for early-stage companies and high-growth startups developing solutions addressing critical issues such as water conservation, efficient irrigation, sustainable agriculture, and waste reduction. By investing in these areas, VC and PE firms help accelerate the development and commercialization of groundbreaking technologies, enabling their widespread adoption and impact [214, 215].

8.4.2 Green bonds and sustainability-linked loans

Green bonds and sustainability-linked loans are innovative financial instruments used to fund projects with positive environmental impacts. Green bonds raise capital exclusively for sustainable initiatives, ensuring transparency and accountability through strict reporting standards. Similarly, sustainability-linked loans offer favorable terms based on borrowers' achievements of predetermined environmental, social, and governance (ESG) targets. Both instruments are increasingly popular in financing projects within the water-food nexus, driving investments in water conservation, sustainable agriculture, and resource efficiency [14].

8.4.2.1 Mechanisms and benefits of green bonds

Green bonds are debt instruments specifically earmarked to raise funds for projects with positive environmental benefits. These bonds provide crucial capital for initiatives that promote sustainability and environmental protection, including those within the water-food nexus. Projects funded by green bonds often focus on water conservation, sustainable agriculture, renewable energy, and pollution reduction. The issuance of green bonds follows rigorous standards and frameworks, such as the Green Bond Principles and Climate Bonds Standard, which ensure that the funds are used appropriately and that the environmental impact is measurable and verifiable. These standards enhance transparency and accountability, making green bonds an attractive option for investors increasingly seeking to align their portfolios with environmental, social, and governance (ESG) criteria. Additionally, green bonds often come with reporting requirements, where issuers must regularly update investors on the progress and outcomes of the funded projects. This level of transparency builds investor confidence and encourages more substantial investment flows into sustainable projects, ultimately contributing to the broader goal of a circular economy [216–219].

8.4.3 Corporate social responsibility and impact investing

CSR and impact investing are pivotal in promoting sustainability within the water-food nexus. CSR involves companies committing to ethical practices and contributing positively to society and the environment, often through initiatives that enhance water efficiency and sustainable agriculture. Impact investing, on the other hand, focuses on generating measurable social and environmental benefits alongside financial returns. Together, these approaches drive corporate engagement and financial support for innovative solutions that address the critical challenges of water and food resource management.

8.4.3.1 Definition and importance of CSR in the water-food nexus

CSR refers to a company's commitment to operating in an economically, socially, and environmentally sustainable manner. In the context of the water-food nexus, CSR initiatives focus on reducing the environmental impact of business operations, promoting sustainable resource management, and contributing to community development. CSR is essential for fostering a culture of sustainability within the corporate sector, encouraging companies to take responsibility for their environmental footprint and invest in practices that benefit society and the environment [217].

8.4.4 Impact investing

Impact investing involves making investments to generate measurable social and environmental impact alongside a financial return. This approach to investing aligns with the goals of the circular economy, particularly within the water-food nexus, where sustainable practices and technologies can have significant positive impacts on communities and the environment.

8.4.4.1 Role of impact investing in the water-food nexus

Impact investors provide capital to businesses and projects that deliver social and environmental benefits, such as improved water management, sustainable agriculture, and enhanced food security. By prioritizing impact alongside financial returns, impact investing channels resources into initiatives that contribute to the circular economy and address critical issues within the water-food nexus [217].

Case 8.3 Acumen Resilient Agriculture Fund

The Acumen Resilient Agriculture Fund (ARAF), sponsored by Acumen and managed by Acumen Capital Partners, is a $58 million initiative to help smallholder farmers in sub-Saharan Africa adapt to climate change. These farmers, who produce up to 80% of the region's food, often live in extreme poverty and are highly vulnerable to climate events like droughts and floods. ARAF invests in early-stage, high-growth agribusinesses that enhance farmer yields and resilience, addressing challenges such as inadequate resources, limited market access, and poor infrastructure. ARAF employs a unique investment model featuring a first-loss layer funded by the IKEA Foundation and the Green Climate Fund to de-risk investments for participants with lower risk tolerance. Additionally, a $5 million Technical Assistance Facility supports portfolio companies experimenting with new technologies and outreach strategies including farmer training and gender-focused initiatives. By deploying risk-tolerant, blended capital and prioritizing the needs of smallholder farmers, ARAF aims to build a sustainable ecosystem of agribusinesses, increase farmers' incomes, and bolster climate resilience in the agriculture sector. The fund is backed by a coalition of organizations including the Green Climate Fund, FMO, and the IKEA Foundation [220].

8.5 Innovative financial instruments

Innovative financial instruments, including blended finance, crowdfunding, and pay-for-performance models, are crucial for advancing the circular economy within the water-food nexus by providing diverse and flexible funding options. These instruments enable capital mobilization from various sources to support sustainable projects. By leveraging different financial mechanisms, these instruments help overcome traditional financing barriers, foster innovation, and promote the scalable adoption of sustainable practices and technologies in the water-food sector.

8.5.1 Blended finance

Blended finance combines public, private, and philanthropic capital to fund projects that deliver social and environmental benefits while achieving financial returns. This innovative financial mechanism is particularly effective in addressing funding gaps for sustainable development projects within the water-food nexus.

8.5.1.1 Concept and structure of blended finance

Blended finance involves using concessional funds from public or philanthropic sources to de-risk investments and attract private sector capital into projects that might otherwise be considered too risky or unprofitable. The structure typically includes several layers of financing, each with different risk and return profiles, enabling the alignment of diverse investor interests. The public or philanthropic capital often takes on the higher-risk tranche, absorbing initial losses and thus making the project more attractive to private investors who take on the lower-risk tranches.

For instance, a typical blended finance structure might include grants or concessional loans from development agencies or foundations to cover initial project costs or provide guarantees. These concessional funds lower the financial risk, encouraging private investors such as commercial banks, venture capitalists, or institutional investors to contribute additional capital. This layered approach leverages more significant amounts of private investment and ensures that projects with substantial social and environmental impacts receive the necessary funding to proceed [217].

8.5.2 Crowdfunding and community-based financing

Crowdfunding and community-based financing have emerged as innovative solutions to fund projects in the water-food nexus. These methods leverage the collective financial power of individuals and communities to support sustainable initiatives. They provide alternative sources of capital that can complement traditional financing mechanisms.

8.5.2.1 Overview of crowdfunding platforms and models
Crowdfunding involves raising small amounts of money from many people, typically through online platforms. There are several models of crowdfunding:
- *Donation-based crowdfunding*: Contributors donate money to support a project without expecting any financial return. This model is often used for community projects or social causes such as building a communal water well or supporting a local sustainable agriculture initiative.
- *Reward-based crowdfunding*: Backers contribute funds in exchange for nonmonetary rewards such as products, services, or recognition. For instance, a startup developing a water-saving irrigation system might offer its supporters early access to the technology or exclusive updates.
- *Equity crowdfunding*: Investors receive equity or shares in the company in return for their financial contributions. This model suits startups and businesses in the water-food sector looking to raise capital while sharing future profits with their investors.
- *Debt crowdfunding (peer-to-peer lending)*: Individuals lend money to a project or business and receive interest payments over time. This model can provide loans for small-scale farmers or water management projects that may not qualify for traditional bank loans [14, 217].

8.5.3 Pay-for-performance and outcome-based financing

Pay-for-performance and outcome-based financing are innovative approaches that link financial incentives to achieving specific, measurable outcomes. These mechanisms are increasingly being applied in the water-food nexus to promote efficiency, sustainability, and accountability.

8.5.3.1 Mechanisms and applications of outcome-based financing
Outcome-based financing involves structuring payments around the achievement of predefined results rather than the mere execution of activities. This approach ensures that funds are only disbursed when specific targets or outcomes are met, fostering greater accountability and performance:

- *Social impact bonds (SIBs)*: In SIBs, private investors provide upfront capital for a project. They are repaid by the government or a philanthropic organization only if the project achieves its predetermined outcomes. This model benefits projects with clear, measurable goals, such as improving water quality or increasing agricultural productivity.
- *Development impact bonds (DIBs)*: Similar to SIBs, DIBs involve private investors funding development projects in low– and middle-income countries. Investors are repaid by donors or development agencies based on the project's success in achieving specified development outcomes.
- *Results-based financing (RBF)*: RBF mechanisms tie funding to the delivery of specific results. For instance, agricultural projects might receive payments for achieving targets related to crop yield improvements, water usage reductions, or adopting sustainable practices [221].

Case 8.4 AgriFI Kenya Challenge Fund

The AgriFI Kenya Challenge Fund, managed by Self-Help Africa and Imani Development Limited, was a European Union initiative to enhance smallholder agriculture through EUR 18 million in financial support to agri-enterprises. Cofunded by SlovakAid and supported by the EIB, the fund provided long-term local currency financing to Equity Bank (Kenya) Limited for on-lending to eligible food and agriculture sector projects. The primary objective of the Challenge Fund was to improve the capacity of 100,000 smallholder farmers and pastoralists to practice environmentally sustainable and climate-smart agriculture as a business in inclusive value chains. By supporting and enabling at least 50 agri-enterprises to increase their turnover by at least 25%, the fund aimed to boost incomes and food security for smallholders and create 10,000 jobs. Agri-enterprises were required to provide match funding, which could come from internal resources, external finance, or coapplicants. Funding decisions were based on economic viability, environmental impact, social impact, and additionality criteria. The program, running from 2018 to 2022, aimed to cultivate 20,000 ha of land using climate-smart approaches and create opportunities in arid and semiarid regions. The AgriFI Kenya Challenge Fund exemplified a targeted effort to support smallholder farmers, enhance sustainable agriculture, and foster inclusive value chains in Kenya [222].

8.6 Strategic partnerships for financing

Strategic partnerships are essential for mobilizing resources and expertise to address the complex challenges of the water-food nexus. By bringing together diverse stakeholders, these collaborations can enhance the scale, impact, and sustainability of financing efforts.

8.6.1 Importance of collaborations and partnerships in financing the water-food nexus

Collaborations and partnerships are vital in financing the water-food nexus because they combine the strengths of various sectors, including government, private industry, nonprofits, and international organizations. These partnerships leverage different types of capital, knowledge, and capabilities to implement comprehensive solutions that single entities might struggle to achieve independently. By aligning the interests and resources of multiple stakeholders, strategic partnerships can address funding gaps, share risks, and promote innovative approaches to sustainable water and food management.

8.6.1.1 Benefits of strategic partnerships

The benefits of strategic partnerships in financing the water-food nexus are numerous, significantly enhancing the success and sustainability of projects:

- *Resource mobilization*: Partnerships can pool financial, human, and technical resources from various sectors, increasing the overall capacity to fund and implement projects.
- *Risk sharing*: Collaborations allow partners to distribute risks, making it easier to undertake large-scale or high-risk projects that would be too challenging for individual entities.
- *Innovation and expertise*: Strategic partnerships bring together diverse perspectives and expertise, fostering innovation and the development of more effective solutions.
- *Increased impact*: By working together, partners can achieve greater scale and impact, ensuring that projects benefit more people and ecosystems.

8.6.2 Strategies for building effective partnerships

Strategies for building effective partnerships are essential for ensuring the success and sustainability of collaborative efforts in the water-food nexus:

- *Identify common goals*: Successful partnerships begin with clearly defined shared objectives. Partners must agree on the desired outcomes and the means to achieve them.
- *Leverage complementary strengths*: Each partner should bring unique strengths to the table. Leveraging these complementary capabilities ensures that the partnership can effectively address various aspects of the project.
- *Establish clear roles and responsibilities*: Clearly defining each partner's roles and responsibilities helps prevent misunderstandings and ensures accountability.

- *Foster open communication*: Regular and transparent communication is crucial for maintaining trust and coordination among partners. Establishing open channels for dialog helps address issues promptly and keeps the project on track.
- *Develop flexible agreements*: Flexibility in agreements allows partners to adapt to changing circumstances and new opportunities. This adaptability is essential for long-term success.

8.6.2.1 Best practices for fostering partnerships

Best practices for fostering partnerships are crucial for enhancing collaboration and achieving long-term success in the water-food nexus:

- *Engage stakeholders early*: Involving all relevant stakeholders from the beginning helps build consensus and commitment to the partnership's goals.
- *Focus on long-term relationships*: Building strong, long-term relationships rather than short-term collaborations ensures sustained impact and continued resource mobilization.
- *Measure and share successes*: Regularly measuring and sharing the partnership's successes can help maintain momentum and attract additional support.
- *Provide capacity building*: Investing in partner capacity building, particularly in developing regions, ensures that all stakeholders can contribute effectively.
- *Celebrate milestones*: Celebrating achievements and milestones helps reinforce the partnership's progress and keeps all partners motivated [14, 15, 217].

Case 8.5 Nestlé and AWS partnership in agricultural water management

Nestlé, in collaboration with the Alliance for Water Stewardship (AWS), is implementing advanced agricultural water management practices in the Middle East and North Africa region to address water scarcity and improve sustainability. The projects focus on enhancing water quantity and quality through drip irrigation, water infrastructure projects, and canal restoration. Nestlé uses excess water to irrigate organic agriculture in Jordan, showcasing sustainable farming in arid regions. In Pakistan, upgraded irrigation systems reduce groundwater consumption and prevent pesticide and fertilizer runoff, improving water efficiency and agricultural productivity. Reforestation and wetland restoration projects also help prevent runoff, enhance groundwater recharge, and support local ecosystems. In Lebanon's Shouf Biosphere Reserve, partnerships have enhanced groundwater reservoir recharge, benefiting local agriculture. Nestlé's efforts include cleaning up and restoring canals, managing wastewater, and educating on water protection to ensure long-term water sustainability. By 2025, Nestlé aims to have all its water sites certified by AWS and lead projects enabling local watersheds to capture as much water as used in their bottling operations, achieving a volumetric water benefit of 4 million m^3 by the end of 2023. These initiatives demonstrate a commitment to regenerating local water cycles to benefit nature and communities [223].

8.7 Challenges and opportunities

The financing of the circular economy in the water-food nexus faces several challenges and also presents numerous opportunities for scaling up investments and fostering sustainable development. Understanding these barriers and opportunities is crucial for stakeholders looking to promote sustainable practices in this sector.

8.7.1 Barriers to financing the circular economy and the water-food nexus

Barriers to financing the circular economy in the water-food nexus present significant challenges that hinder the flow of capital into sustainable projects. Key obstacles include regulatory and policy inconsistencies, market and financial risks, and unreliable data and metrics. These factors create an uncertain environment for investors, making it difficult to assess potential returns and limiting the scalability of innovative solutions in this critical sector.

8.7.1.1 Regulatory and policy challenges

Regulatory and policy frameworks can significantly impact the flow of financing into the water-food nexus. Inconsistent regulations, lack of clear policies, and bureaucratic hurdles often create an uncertain environment for investors. For example, differing water usage rights and agricultural policies across regions can complicate investment decisions and limit the scalability of successful projects. Additionally, subsidies for traditional agricultural practices discourage investment in innovative, sustainable solutions.

8.7.1.2 Market and financial risks

Investing in the water-food nexus involves inherent market and financial risks. The high upfront costs and long payback periods of many sustainable technologies can deter investors from seeking quicker returns. Market volatility, fluctuating commodity prices, and economic instability increase the perceived risk of investing in this sector. Furthermore, there is often a lack of reliable data and metrics to evaluate the performance and impact of circular economy projects, making it difficult for investors to assess potential returns accurately.

8.7.2 Opportunities for scaling up investments

Opportunities for scaling up investments in the water-food nexus are emerging through innovative financial instruments and models. Green bonds, sustainability-linked loans, blended finance, crowdfunding, and outcome-based financing offer new avenues for funding sustainable projects. These innovations attract diverse investors, de-risk investments, and ensure accountability, making it possible to mobilize significant capital for advancing the circular economy in the water-food sector.

8.7.2.1 Emerging trends and innovations in financing the water-food nexus

Despite these challenges, several emerging trends and innovations offer opportunities to scale up investments in the water-food nexus. The increasing availability of green bonds and sustainability-linked loans provides new avenues for funding projects that promote environmental sustainability. These financial instruments are designed to attract a broad range of investors by linking financial returns to achieving specific environmental outcomes.

Innovative financing models such as blended finance, crowdfunding, and outcome-based financing present significant opportunities. Blended finance combines public, private, and philanthropic funds to de-risk investments and attract private capital. Crowdfunding platforms enable small investors to contribute to sustainable projects, while outcome-based financing ties payments to achieving predefined results, ensuring accountability and performance [14, 15, 217].

8.7.2.2 Recommendations for policymakers and investors

To capitalize on these opportunities and address the existing barriers, policymakers and investors must adopt several strategies.

For policymakers, creating a supportive regulatory environment is essential. This includes harmonizing regulations across regions, providing clear and consistent policy frameworks, and removing bureaucratic obstacles that hinder investment. Policymakers should also consider redirecting subsidies from traditional practices to support sustainable technologies and practices within the water-food nexus.

Additionally, developing standardized metrics and data collection methods to evaluate the performance and impact of circular economy projects can provide investors with the information they need to make informed decisions. Encouraging transparency and accountability through rigorous reporting standards can further build investor confidence.

For investors, diversifying investment portfolios to include a mix of traditional and innovative financing mechanisms can mitigate risks and enhance returns. Embracing new financial instruments, such as green bonds and sustainability-linked loans, can align investment strategies with sustainability goals. Collaborating with public and philanthropic organizations through blended finance models can leverage additional resources and share risks.

Investors should also seek to engage with local communities and stakeholders to understand the unique challenges and opportunities within the water-food nexus. Building strong relationships with local partners can facilitate project implementation and ensure investments deliver meaningful, sustainable outcomes.

By addressing regulatory and policy challenges, mitigating market and financial risks, and leveraging emerging trends and innovations, stakeholders can unlock significant opportunities for scaling up investments in the water-food nexus. These efforts will contribute to the transition toward a circular economy and enhance food security and water sustainability for future generations [14, 15, 217].

Chapter 9
Best practices and conclusion

Abstract: This chapter explores the best practices for implementing a circular economy within the water-food nexus, focusing on reducing water usage, reusing, and recycling water, nutrient recovery, and ecosystem restoration. Through global case studies, it highlights innovative financial models, technology integration, and collaborative partnerships that support sustainable water and food management. By adopting the 5R approach – reduce, reuse, recycle, recover, and restore – regions can enhance water security, improve agricultural productivity, and foster environmental resilience. This chapter emphasizes the role of ongoing innovation and collaboration in driving long-term sustainability in water and food systems.

9.1 Introduction

The transition to a circular economy, particularly within the water-food nexus, is critical in addressing global challenges such as water scarcity, food insecurity, and environmental degradation. This chapter focuses on best practices that regions and sectors can adopt to move from a linear economic model toward a circular approach. Implementing strategies that optimize resource use – such as reducing, reusing, and recycling water and recovering valuable nutrients – can contribute to sustainable water and food management. The chapter also highlights the importance of innovation, technology, and financial collaboration in supporting the shift toward more resilient and sustainable systems.

9.2 Defining the circular economy and the water-food nexus and the 5Rs

Based on the case studies in Chapter 2: Defining the Circular Economy and the Water-Food Nexus, the following best practices have been identified for regions worldwide aiming to transition from a linear to a circular economic model within the water-food nexus. These best practices focus on reducing waste, optimizing resource use, and fostering sustainable water and food management through the circular economy's 5R approach – reduce, reuse, recycle, recover, and restore. By implementing these practices, regions can enhance water security, improve agricultural productivity, and contribute to environmental resilience in the face of growing global challenges.

https://doi.org/10.1515/9783111341385-009

9.2.1 Reducing water usage

The following best practices have been identified from the case studies focused on reducing water usage in agriculture.

9.2.1.1 Tiered incentives for water-efficient irrigation

Water management authorities can introduce tiered financial incentives to encourage the adoption of advanced irrigation technologies, such as drip and sprinkler systems, that reduce water waste. The Arizona Water Irrigation Efficiency Program demonstrates the success of providing direct financial support to farmers for purchasing efficient irrigation systems, resulting in significant water savings.

9.2.1.2 Drought resilience grants

Governments can implement innovation grants targeting drought resilience in agriculture, funding projects that conserve water and improve soil moisture retention. Australia's Drought Resilience Innovation Grants highlight the effectiveness of funding for slow-release fertilizers and groundwater reuse technologies to support long-term water conservation in drought-prone regions.

9.2.2 Reusing water

As demonstrated by several case studies, best practices for water reuse include the following.

9.2.2.1 Gray water reuse for agriculture

In areas with water scarcity, gray water reuse can be a practical and cost-effective solution for irrigation in small-scale agriculture and gardens. South Africa's guidelines for gray water reuse provide a model for other regions to implement, allowing communities to conserve freshwater while promoting sustainable agricultural practices.

9.2.2.2 Industrial water reuse

Regions with large-scale industrial activities can explore nonconventional water sources, such as treated wastewater from oil and gas fields, for agricultural and industrial applications. The Abu Dhabi initiative demonstrates how collaboration between industries and environmental agencies can lead to the successful reuse of produced water, reducing reliance on freshwater resources.

9.2.3 Recycling water

The following best practices can be implemented from the case studies on recycling treated water.

9.2.3.1 Agricultural use of recycled wastewater

As seen in Kent County, Maryland, municipal wastewater can be treated and reused for agricultural irrigation. This practice addresses water scarcity and provides a reliable and regulated water source for farmers, enhancing crop yields even in areas with unpredictable rainfall.

9.2.3.2 Urban recycled water for agricultural irrigation

Wastewater treatment plants can prioritize recycled water production for urban landscaping and agricultural irrigation, as demonstrated by the Al Wathbah-2 Wastewater Treatment Plant in Abu Dhabi. This reduces the need for expensive and energy-intensive desalinated water, promoting a more sustainable approach to water management.

9.2.4 Recovering nutrients from wastewater

Best practices for nutrient recovery in agriculture include the following.

9.2.4.1 Biosolids recycling

Using biosolids treated from sewage sludge provides a sustainable alternative to chemical fertilizers, improving soil health and crop productivity. The biosolids programs at DC Water's Blue Plains and United Utilities in the UK illustrate how biosolids can be safely recycled and used across farms, supporting sustainable agricultural practices while minimizing environmental impacts.

9.2.4.2 Advanced nutrient recovery systems

Water treatment plants can implement advanced technologies, such as thermal hydrolysis and anaerobic digestion, to treat biosolids, ensuring that they meet high safety and quality standards for agricultural use. This recycles nutrients like nitrogen and phosphorus and contributes to soil carbon sequestration efforts.

9.2.5 Restoring ecosystems through nature-based solutions

The following best practices can be implemented to restore natural water systems and enhance agricultural sustainability.

9.2.5.1 Grants and incentives for conservation practices

Governments can offer grants and financial incentives to farmers to adopt best management practices (BMPs) that conserve water, reduce soil erosion, and protect water quality. The Maryland Agricultural Water Quality Cost-Share Program provides a model for other regions, offering financial support for cover cropping, buffer planting, and other conservation measures.

9.2.5.2 Collaborative water quality improvement projects

Local, farmer-led projects to improve water quality through sustainable farming practices can be highly effective. The Mulkear Operational Group project in Ireland demonstrates how collaborative efforts between farmers and water management authorities can lead to innovative, practical solutions for water conservation and improved agricultural practices, rewarded through results-based payments.

9.3 Enhancing water efficiency and conservation

Based on the case studies in Chapter 3: Enhancing Water Efficiency and Conservation Through Circular Agriculture Solutions, the following best practices for regions worldwide have been identified aiming to implement sustainable water management and conservation practices in agriculture. These best practices focus on optimizing water usage, enhancing irrigation efficiency, and leveraging technology and policy frameworks to ensure the long-term viability of water resources in agriculture.

9.3.1 Enhancing water efficiency and conservation

Several critical best practices have emerged from the case studies focused on enhancing water efficiency and conservation.

9.3.1.1 Crop diversification and climate resilience

Encouraging farmers to diversify crops, especially by integrating drought-resistant species such as lucerne, can significantly improve forage production and farm resilience against climate change. Slovenia's project demonstrates how using crop rotation strategies, particularly with legumes, improves soil fertility, reduces reliance on external fertilizers, and enhances water efficiency while maintaining agricultural productivity.

9.3.1.2 Precision irrigation through cloud-based systems

Utilizing cloud-based water management systems, like the COALA project in Australia, can optimize irrigation practices by providing real-time data on water usage, weather patterns, and other variables. This technology improves irrigation efficiency, reduces water waste, and enhances agricultural sustainability, achieving measurable outcomes such as a 20% improvement in water use efficiency.

9.3.2 Implementing rainwater harvesting and artificial recharge

Best practices for regions seeking to enhance water availability through nonconventional water sources include the following.

9.3.2.1 Rainwater harvesting and artificial recharge

Capturing and storing rainwater, particularly in regions highly dependent on monsoon rains, offers a reliable water source for irrigation and improves groundwater levels. India's success with rainwater harvesting and artificial groundwater recharge demonstrates how these practices can significantly increase cropping intensity and ensure agricultural sustainability in water-scarce areas.

9.3.3 Leveraging ICT and AI for water management

The following best practices have been identified from the case studies integrating technology into water management.

9.3.3.1 Integration of ICT and AI for real-time water management

Deploying soil sensors and leveraging AI for real-time data collection, as Weenat's Europe-wide network demonstrated, allows for precise soil moisture level monitoring. This approach enhances water conservation by ensuring that irrigation is applied only when and where necessary, significantly reducing water consumption in agriculture. The project's success in saving millions of cubic meters of water underscores the importance of technology in addressing water scarcity challenges.

9.3.4 Advancing evapotranspiration management

Effective evapotranspiration (ET) management is critical to improving water use efficiency. Best practices identified from the case studies include the following.

9.3.4.1 Advanced soil moisture monitoring and ET management

Systems like the Soil Moisture Integration and Prediction System in Australia that integrate satellite imagery, ground-based soil data, and climate information to provide daily estimates of soil moisture at high resolution are instrumental in optimizing irrigation schedules. This precise data allows farmers to match water usage with crop needs, reducing wastage and enhancing agricultural sustainability.

9.3.5 Implementing water trading systems

Water trading systems offer flexibility in water usage and promote efficient allocation of resources. Essential best practices include the following.

9.3.5.1 Water trading for efficient resource allocation

The water trading system in Australia's Murray-Darling Basin provides an efficient way to allocate water resources based on market demand. By enabling the buying and selling of water rights, this system ensures that water is directed to high-value agricultural uses, especially during times of scarcity. The flexibility of this approach reduces financial risks for farmers and supports sustainable water management across large agricultural regions.

9.3.6 Promoting financial incentives and irrigation upgrades

Best practices for enhancing irrigation efficiency include the following.

9.3.6.1 Financial incentives for irrigation efficiency

Programs like Washington's Irrigation Efficiencies Grant Program, which provides financial support for farmers to install more efficient irrigation systems, highlight the importance of incentivizing sustainable water use. These initiatives improve water use efficiency and restore natural water flows, supporting agricultural productivity and environmental conservation.

9.3.6.2 Agricultural water management requirements

In California, agricultural water management plans mandated by the Water Conservation Act ensure that large-scale water suppliers adopt efficient water management practices. This regulatory framework can help regions balance agricultural productivity with sustainable water use, reducing the sector's water footprint.

9.3.7 Education and awareness for water conservation

Education and awareness play a crucial role in fostering sustainable water management practices. Essential best practices include the following.

9.3.7.1 Educational programs for conservation

Programs like North Carolina State University's Soil and Water Resource Conservation Workshop demonstrate the importance of educating the next generation on natural resource conservation. Such programs raise awareness about water management, soil health, and sustainability, encouraging future leaders to pursue careers in environmental conservation and resource management.

9.4 Sustainable agriculture and nature-based solutions: preserving water quality

Based on the case studies in Chapter 4: Sustainable Agriculture and Nature-Based Solutions Preserving Water Quality, the following best practices have been identified for regions worldwide aiming to implement sustainable agriculture practices and nature-based solutions to preserve water quality. These best practices focus on reducing nutrient runoff, improving soil health, and integrating ecofriendly strategies into agriculture to enhance productivity and environmental quality.

9.4.1 Implementing climate-smart and sustainable agriculture

Several critical best practices have emerged from the case studies focusing on climate-smart agriculture and sustainable practices.

9.4.1.1 Climate-smart agriculture and innovation grants

Programs like Australia's Climate-Smart Agriculture Program promote the integration of sustainable resource management practices that enhance climate resilience and productivity. Innovation and capacity-building grants encourage farmers to adopt sustainable practices, protect natural resources, and improve biodiversity while aligning agricultural activities with global sustainability goals.

9.4.1.2 Sustainable agriculture demonstration farms

Establishing community demonstration farms, as seen in Ghana, is an effective way to disseminate knowledge on sustainable crop production and water conservation techniques. These educational hubs promote drought-tolerant crops, sustainable farming practices, and efficient water use, contributing to improved water quality and agricultural sustainability.

9.4.2 Integrating nature-based solutions for water quality

Best practices for regions aiming to enhance water quality through nature-based solutions include the following.

9.4.2.1 Constructed wetlands for runoff treatment

Resources such as the Farm Wetland Sizing Toolkit, developed by the Wildfowl and Wetlands Trust, enable farmers to estimate the size and design of constructed wetlands for treating farm runoff. Constructed wetlands reduce nutrient pollution and improve water quality by acting as natural filtration systems, aligning with broader environmental conservation goals.

9.4.2.2 Agroforestry and mixed farming systems

Projects like AGROMIX in Europe demonstrate the value of integrating agroforestry and mixed farming systems to enhance soil health, reduce runoff, and create more resilient agricultural landscapes. By fostering synergies between these systems farmers can improve efficiency and reduce environmental impacts, contributing to low-carbon, climate-resilient farming practices.

9.4.3 Fostering innovation and knowledge sharing

From the case studies promoting collaboration and knowledge dissemination, the following best practices have been identified.

9.4.3.1 Participatory research and innovation

Engaging farmers and stakeholders in participatory research, as seen in the AGROMIX project, helps codesign innovative agroecological solutions tailored to local conditions.

This approach encourages the development of practical toolkits and actionable strategies that foster sustainable agricultural transitions.

9.4.3.2 Capacity building and education initiatives

Programs that focus on building community capacity, like the Climate-Smart Agriculture Program in Australia, are critical for spreading knowledge about sustainable agriculture and natural resource management. These initiatives empower communities to implement eco-friendly practices that contribute to agricultural productivity and the preservation of water quality.

9.5 Integrated water and nutrient reuse and recycling in sustainable agriculture

Based on the case studies in Chapter 5: Integrated Water and Nutrient Reuse and Recycling in Sustainable Agriculture, the following best practices have been identified for regions aiming to implement sustainable water reuse and nutrient recycling systems in agriculture. These best practices focus on optimizing water and nutrient recovery, enhancing resource efficiency, and leveraging advanced technologies to support ecosystem resilience and agricultural productivity.

9.5.1 Water reuse and recycling in agriculture

From the case studies on water reuse and recycling in agriculture, the following key best practices have emerged.

9.5.1.1 High-quality recycled water for agriculture

Australia's Northern Adelaide Irrigation Scheme (NAIS) demonstrates the importance of integrating advanced water recycling systems to provide high-quality, climate-independent water for agricultural use. The scheme supports horticultural activities, fruit orchards, and other high-value crops, fostering agricultural productivity and economic growth. By providing reliable recycled water, NAIS promotes efficient water use and ensures long-term sustainability in agriculture.

9.5.1.2 Rainwater harvesting for drip irrigation

In Jamaica, integrating rainwater harvesting with drip irrigation systems offers a reliable and sustainable water supply for farmers facing irregular rainfall patterns. This practice, which applies water directly to the root zones of crops, maximizes water use efficiency, and improves crop yields. The combination of rainwater harvesting and drip irrigation helps conserve groundwater resources and reduces the environmental impacts of traditional irrigation methods.

9.5.2 Nutrient recycling and resource efficiency

Best practices for nutrient recycling in sustainable agriculture include the following.

9.5.2.1 Phosphorus recovery through struvite crystallization

Waternet's Airprex process at the Amsterdam West Waste Water Treatment Plant showcases the potential of recovering valuable nutrients like phosphorus from wastewater. By crystallizing struvite, Waternet resolved operational issues such as scaling and produced high-quality fertilizer. This closed-loop system supports sustainable agriculture by reducing the reliance on synthetic fertilizers and promoting efficient nutrient use.

9.5.2.2 Enhanced resource efficiency in wastewater treatment

The Airprex process also demonstrates how integrating nutrient recovery systems into wastewater treatment can improve operational efficiency. This approach offers significant economic and environmental benefits by reducing phosphate levels in the wastewater, enhancing sludge dewaterability, and lowering chemical usage, making it a replicable model for other wastewater treatment plants worldwide.

9.5.3 Implementing advanced technologies for sustainable water and nutrient management

The following best practices have been identified from the case studies focusing on advanced water and nutrient management technologies.

9.5.3.1 Advanced water recycling infrastructure

Large-scale water recycling systems, like NAIS in Australia, can provide a consistent water supply to agricultural regions. By investing in infrastructure such as wastewater treatment plants, seasonal storage, and distribution networks, regions can ensure the long-term viability of water resources for agriculture, even in the face of climate variability.

9.5.3.2 Integration of drip irrigation systems with rainwater harvesting

Drip irrigation systems, particularly when integrated with rainwater harvesting, can significantly enhance water use efficiency in agriculture. This approach, as seen in

Jamaica, ensures that crops receive precise water application, reducing water loss through evaporation or percolation and improving the efficiency of fertilizer application through fertigation.

9.5.4 Economic and environmental benefits of water and nutrient recycling

Best practices for promoting the economic and environmental benefits of integrated water and nutrient recycling include the following.

9.5.4.1 Economic gains from water and nutrient recycling

Initiatives like Waternet's Airprex process show how wastewater treatment can produce valuable byproducts such as struvite, a high-quality fertilizer. The economic benefits of nutrient recovery through processes like Airprex, with reduced costs in treatment and operational efficiency, contribute to long-term financial sustainability.

9.5.4.2 Environmental benefits of sustainable water and nutrient management

Reducing groundwater extraction, preventing soil degradation, and minimizing nutrient runoff are critical environmental outcomes of integrating water reuse and nutrient recycling systems. As demonstrated by NAIS in Australia and the rainwater-fed drip irrigation systems in Jamaica, these practices contribute to preserving water sources, reducing pollution, and enhancing soil health.

9.6 Sustainable food systems advancing water conservation and energy efficiency

Based on the case studies in Chapter 6: Sustainable Food Systems Advancing Water Conservation and Energy Efficiency, the following best practices have been identified for regions aiming to implement innovations in food systems that enhance water conservation and energy efficiency. These best practices focus on leveraging advanced farming technologies, promoting circular waste management, and integrating renewable energy sources to support sustainable food production while minimizing environmental impacts.

9.6.1 Advancing water conservation in sustainable food systems

The following best practices have emerged from the case studies focused on improving water conservation in food systems.

9.6.1.1 Hydroponic farming for water efficiency
Bustanica's hydroponic vertical farm in Dubai showcases the potential of water-efficient farming technologies. By using a closed-loop system that recirculates water and captures evaporated water for reuse, Bustanica has achieved a 95% reduction in water use compared to traditional farming methods. This approach, which conserves approximately 250 million liters of water annually, exemplifies how water-efficient hydroponics can contribute to sustainable food production, particularly in regions facing water scarcity.

9.6.1.2 Integrating drip irrigation with controlled environments
GigaFarm in the UAE integrates advanced vertical farming technologies with controlled environments, significantly reducing water usage and fertilizer consumption. By optimizing water efficiency through innovative systems and recycling food scraps into compost and clean water, GigaFarm demonstrates a sustainable, circular approach to food production that can be replicated in other arid regions facing similar challenges.

9.6.2 Leveraging genetic engineering and selective breeding for climate-resilient crops

Best practices for enhancing the resilience of crops through genetic advancements include the following.

9.6.2.1 Developing climate-resilient crops
The Ancient Environmental Genomics Initiative for Sustainability (AEGIS) focuses on utilizing the genetic diversity found in ancient environmental DNA to develop crops resilient to modern climate challenges. By leveraging advanced DNA sequencing and bioinformatics tools, AEGIS aims to enhance global food security by creating crops that can withstand extreme weather conditions, showcasing the potential of genetic innovation in sustainable agriculture.

9.6.3 Reducing food waste through circular initiatives

The following best practices have been identified from the case studies on food waste reduction.

9.6.3.1 Community-driven food waste reduction

Love Food Hate Waste NZ provides a model for reducing food waste through community engagement and education. By encouraging individuals to plan their meals, store food properly, and minimize waste at its source, this initiative helps households reduce their carbon footprint and promotes a culture of sustainability. The program's focus on changing consumption habits to reduce waste aligns with global sustainability goals.

9.6.3.2 Organic waste recycling for energy and fertilizer production

The Rialto Bioenergy Facility in California demonstrates how anaerobic digestion can convert organic waste into renewable biopower and marketable fertilizer. This process helps reduce landfill waste, lower greenhouse gas emissions, and produce valuable agricultural byproducts like fertilizer pellets, making it an effective strategy for managing food waste sustainably.

9.6.4 Integrating renewable energy in agriculture

Best practices for promoting energy efficiency and renewable energy use in agriculture include the following.

9.6.4.1 Renewable energy integration for off-grid farming

Botswana's National Development Bank program for small and medium-sized off-grid farmers supports transitioning from diesel-powered generators to solar energy and water-efficient irrigation practices. By replacing diesel engines with solar power and promoting technologies like drip irrigation and hydroponics, this initiative enhances agricultural productivity while reducing greenhouse gas emissions, making it a replicable model for other regions seeking to reduce their reliance on fossil fuels.

9.6.4.2 Biogas production from organic waste

The Rialto Bioenergy Facility's use of anaerobic digestion to produce biogas from organic waste offers a sustainable solution for generating renewable energy. This approach reduces the environmental impact of organic waste while providing a clean energy source for local communities, supporting sustainable agriculture and energy efficiency.

9.7 Urban agriculture and water management circular solutions for urban landscapes

Based on the case studies in Chapter 7: Urban Agriculture and Water Management Circular Solutions for Urban Landscapes, the following best practices have been identified for regions aiming to implement circular urban agriculture systems that enhance water management, food security, and environmental resilience. These best practices focus on integrating urban agriculture into city environments, utilizing green infrastructure, and promoting waste-to-resource solutions for sustainable urban ecosystems.

9.7.1 Advancing sustainable water use in urban agriculture

The following best practices have emerged from the case studies focusing on sustainable water use in urban agriculture.

9.7.1.1 Rainwater harvesting for urban agriculture

Reflo's projects in Milwaukee highlight the importance of integrating rainwater harvesting systems into urban agriculture. Reflo's initiatives ensure a reliable, sustainable water source for community gardens and urban farms by constructing underground cisterns that collect and store stormwater for irrigation. This approach also incorporates green infrastructure elements like bioswales, which filter and manage stormwater, further enhancing water efficiency and sustainability in urban agriculture.

9.7.1.2 Closed-loop water systems for urban farming

Utilizing closed-loop water systems, like those implemented at Cream City Farms, ensures that water collected from urban landscapes is efficiently reused for irrigation. Combined with solar-powered pumps, these systems demonstrate how urban agriculture can reduce reliance on municipal water supplies and increase resilience to water shortages.

9.7.2 Transforming waste into resources for sustainable urban systems

Best practices for integrating waste-to-resource initiatives in urban agriculture include the following.

9.7.2.1 Food waste to biogas conversion

Stockholm's food waste initiative showcases the potential of transforming food waste into biogas for energy. By mandating the separation of food waste from other waste, Stockholm maximizes resource utilization and reduces greenhouse gas emissions. This initiative provides a model for cities aiming to integrate circular waste management practices into urban food systems, converting organic waste into renewable energy and reducing the environmental footprint of food production.

9.7.2.2 Nutrient cycling through urban farms and food forests

San Antonio's food forests and urban farms, as explored in the "Vibrant Land" study, demonstrate the benefits of using urban agriculture to cycle nutrients, reduce urban heat, and retain floodwaters. Food forests contribute to carbon sequestration and green space access while addressing food insecurity in vulnerable communities. This nature-based approach to urban farming provides a blueprint for cities looking to enhance both food security and environmental sustainability.

9.7.3 Supporting urban agriculture through grants and community engagement

The following best practices have been identified from the case studies focused on funding and supporting urban agriculture.

9.7.3.1 Grant programs for urban agriculture infrastructure

The DC Urban Agriculture Small Award Program, administered by the Chesapeake Bay Trust and the District of Columbia's Department of Energy and Environment, highlights the importance of financial support for expanding urban agriculture. Grants aimed at socially disadvantaged farmers and under-resourced communities enable the growth of urban farms, improving access to fresh food and promoting sustainable agricultural practices. This model of targeted financial support can be replicated in other cities seeking to address food security and support local farming initiatives.

9.7.3.2 Collaborative planning and community engagement

The study on San Antonio's urban farms emphasizes the value of collaborative planning between government agencies, academic institutions, and community organizations. Engaging communities in the design and management of urban agriculture projects ensures that these initiatives address local needs such as food insecurity, urban cooling, and equitable access to green spaces. This inclusive approach fosters a sense of ownership and long-term sustainability for urban agriculture projects.

9.7.4 Promoting ecosystem services through urban agriculture

Best practices for maximizing the ecosystem services urban agriculture provides include the following.

9.7.4.1 Green infrastructure integration with urban farms

Incorporating green infrastructure, such as bioswales and rain gardens, into urban farms and gardens enhances their ability to manage stormwater, reduce urban heat, and improve air quality. Reflo's projects in Milwaukee exemplify how urban agriculture can be integrated into broader green infrastructure strategies to create multifunctional landscapes supporting food production and environmental health.

9.7.4.2 Food forests for urban resilience

The establishment of food forests, as seen in San Antonio, provides long-term benefits for urban resilience. These forests offer fresh produce and contribute to floodwater retention, nutrient cycling, and carbon sequestration, making them a valuable addition to urban landscapes focused on sustainability and climate adaptation.

9.8 Financing the circular economy investment strategies and partnerships in the water-food nexus

Based on the case studies in Chapter 8: Financing the Circular Economy Investment Strategies and Partnerships in the Water-Food Nexus, the following best practices have been identified for regions aiming to finance and promote the circular economy within the water-food nexus. These best practices focus on leveraging innovative financial instruments, building strategic partnerships, and implementing investment strategies that promote sustainability, resilience, and economic growth in the agricultural and water sectors.

9.8.1 Leveraging financial instruments for sustainable water and food management

Several critical best practices that have emerged from the case studies focused on financing sustainable water and food management.

9.8.1.1 Blended finance and public-private partnerships

The Acumen Resilient Agriculture Fund demonstrates the value of blended finance, combining capital from private investors with risk-tolerant funding from foundations like the IKEA Foundation and the Green Climate Fund. This model enables early-stage investment in high-growth agribusinesses, helping smallholder farmers in sub-Saharan Africa improve their resilience to climate change while attracting private capital by de-risking investments.

9.8.1.2 Competitive grant programs for water efficiency

California's Agriculture Water Use Efficiency Program highlights the importance of public grant initiatives that support water infrastructure improvements. By offering competitive grants to farmers and agricultural agencies, this program fosters the adoption of water-efficient practices, reduces greenhouse gas emissions, and enhances groundwater recharge. The project's financial support for water conveyance upgrades and on-farm efficiency demonstrates a comprehensive approach to sustainable agriculture financing.

9.8.2 Supporting climate adaptation and resilience through strategic partnerships

Best practices for building climate resilience and adaptation strategies through strategic partnerships include the following.

9.8.2.1 Collaborative adaptation projects

The Climate Change Adaptation in Rural Areas of India (CCA-RAI) project, funded by the German Federal Ministry for Economic Cooperation and Development, illustrates the power of international partnerships in building climate resilience. This initiative integrates climate change adaptation into rural development planning and policy while training policymakers and developing localized climate action plans. Collaborative efforts between governments, NGOs, and research institutions can enhance community capacities and improve socioeconomic conditions in vulnerable regions.

9.8.2.2 Water management partnerships for arid regions

The Nestlé and Alliance for Water Stewardship (AWS) partnership exemplifies how private companies can collaborate with organizations to implement water management solutions in regions facing water scarcity. By improving irrigation systems, restoring wetlands, and promoting sustainable farming practices, Nestlé supports long-term water sustainability in the Middle East and North Africa. These partnerships foster responsible water stewardship while also supporting local communities and ecosystems.

9.8.3 Fostering sustainable agricultural enterprises through challenge funds

The case studies focused on funding sustainable agricultural enterprises and identified the following best practices.

9.8.3.1 Challenge funds for climate-smart agriculture

The AgriFI Kenya Challenge Fund demonstrates how challenge funds can be an effective tool for fostering climate-smart agriculture. By providing financial support to agri-enterprises and requiring match funding, the initiative promotes sustainable farming practices, creates jobs, and enhances food security for smallholder farmers. This cofunding model ensures that agricultural enterprises are economically viable while delivering social and environmental benefits.

9.8.3.2 Technical assistance for agricultural enterprises

The Acumen Resilient Agriculture Fund includes a Technical Assistance Facility, providing financial and technical support for portfolio companies to experiment with new technologies and outreach strategies. This approach emphasizes the importance of technical assistance in ensuring that investments lead to innovation and scalability in sustainable agriculture.

9.8.4 Promoting inclusive value chains and local economic growth

Best practices for fostering inclusive value chains and supporting local economic growth in the water-food nexus include the following.

9.8.4.1 Inclusive value chains in agricultural financing

The AgriFI Kenya Challenge Fund creates inclusive value chains by supporting smallholder farmers and pastoralists in arid and semiarid regions. The initiative ensures that even vulnerable populations can benefit from sustainable agriculture as a business by offering long-term local currency financing through partnerships with financial institutions like Equity Bank. This approach highlights the need for financial solutions tailored to local contexts and inclusive value chains.

9.8.4.2 Enhancing economic impact through sustainable water use

California's Agriculture Water Use Efficiency Program demonstrates how investments in water efficiency can yield significant economic benefits, including water savings, greenhouse gas emissions reductions, and improved crop health. By investing in sustainable water infrastructure, the program supports the long-term viability of agriculture in the state while promoting economic growth through job creation and increased productivity.

9.8.5 Scaling investments through public and private collaboration

Best practices for scaling investments in the circular economy include the following.

9.8.5.1 Public-private collaborations for water sustainability

Partnerships like Nestlé and AWS show the importance of combining public and private resources to achieve large-scale water sustainability goals. Nestlé's goal to regenerate local water cycles and achieve a volumetric water benefit of 4 million m^3 by 2023 highlights how companies can play a critical role in sustainable water management while benefiting nature and communities. Collaborative investments in water infrastructure and stewardship can set a precedent for other regions facing similar challenges.

9.9 Conclusion

The transition from a linear to a circular economy, particularly within the water-food nexus, is increasingly recognized as a vital pathway to sustainable development. A circular economy aims to minimize waste and maximize available resources by reducing, reusing, recycling, recovering, and restoring materials. Within the context of the water-food nexus, this approach holds promise for addressing global challenges such as water scarcity, food insecurity, and environmental degradation. As explored throughout this chapter, implementing the best practices identified across various regions and sectors can significantly enhance water and food security while promoting long-term environmental resilience.

The circular economy in the water-food nexus revolves around optimizing resource use at every stage of production and consumption from minimizing water usage in agriculture to recovering nutrients from wastewater. Each of the best practices outlined in this chapter illustrates how different regions and sectors are already advancing toward circular economy principles. These practices demonstrate the tangible benefits of transitioning to a circular economy and provide valuable insights into how other regions and sectors can follow suit.

A fundamental circular economy principle is reducing resource consumption and waste generation. Reducing water usage in agriculture is particularly critical for the water-food nexus, given that agriculture is one of the largest consumers of freshwater resources globally. Best practices, such as tiered financial incentives for water-efficient irrigation systems, as seen in the Arizona Water Irrigation Efficiency Program, show how financial mechanisms can drive the adoption of advanced irrigation technologies like drip and sprinkler systems. These programs can significantly reduce water waste, improve crop yields, and contribute to water conservation efforts by incentivizing farmers to use water more efficiently.

Similarly, innovation grants aimed at improving drought resilience in agriculture, such as Australia's Drought Resilience Innovation Grants, have proven effective in encouraging the development and adoption of technologies that enhance water efficiency. These initiatives enable farmers to conserve water by using slow-release fertilizers and reusing groundwater, contributing to long-term water sustainability in drought-prone areas. Reducing water usage is a crucial pillar of the circular economy in agriculture, as it directly addresses the urgent need for sustainable water management in the face of increasing water scarcity and climate variability.

Water reuse represents another critical aspect of the circular economy in the water-food nexus. As implemented in South Africa, best practices such as gray water reuse for small-scale agriculture provide a practical solution for conserving freshwater resources while promoting sustainable agricultural practices. In regions facing water scarcity, reusing gray water for irrigation reduces the strain on freshwater supplies and enables communities to maintain productive agricultural systems. Similarly, industrial water reuse initiatives, like those in Abu Dhabi, demonstrate how large-scale industries can collaborate with environmental agencies to reduce reliance on freshwater resources by reusing treated wastewater in agricultural and industrial applications. These practices help conserve water and foster collaboration between sectors, highlighting the interconnectedness of water, food, and industry in the circular economy.

Recycling water, mainly treated wastewater, offers further opportunities for sustainable water management within the water-food nexus. Case studies from Kent County, Maryland, and the Al Wathbah-2 Wastewater Treatment Plant in Abu Dhabi showcase how treated wastewater can be safely and effectively reused for agricultural irrigation and urban landscaping. By prioritizing recycled water production, these regions reduce the need for expensive and energy-intensive desalinated water, promoting a more sustainable approach to water management. Additionally, recycling water supports the goals of the circular economy by closing the loop on water use, ensuring that valuable water resources are reused rather than wasted.

Nutrient recovery from wastewater is another essential component of the circular economy in the water-food nexus. Nutrient recovery systems, such as those implemented at DC Water's Blue Plains and United Utilities in the UK, demonstrate how biosolids can be safely recycled and used in agriculture, providing a sustainable alternative to chemical fertilizers. These programs support soil health, enhance crop productivity, and minimize the environmental impacts of traditional agricultural practices. Furthermore, advanced nutrient recovery technologies, such as thermal hydrolysis and anaerobic digestion, enable water treatment plants to treat biosolids to high safety and quality standards, ensuring they can be reused in agriculture while contributing to soil carbon sequestration efforts.

Restoring ecosystems through nature-based solutions is another best practice that aligns with the circular economy principles of the water-food nexus. For conservation practices like Maryland's Agricultural Water Quality Cost-Share Program, grants and incentives encourage farmers to adopt best management practices that conserve water, reduce soil erosion, and protect water quality. These practices benefit the environment and enhance agricultural productivity by maintaining healthy ecosystems. Collaborative water quality improvement projects, such as Ireland's Mulkear Operational Group, further highlight the value of farmer-led initiatives in achieving sustainable water management. By promoting sustainable farming practices and rewarding participants through results-based payments, these projects foster a sense of ownership and responsibility among farmers, leading to long-term improvements in water and food security.

The role of technology in advancing water efficiency and conservation within the water-food nexus cannot be overstated. Precision irrigation systems, such as those developed through Australia's COALA project, demonstrate how cloud-based water management systems can optimize irrigation practices by providing real-time data on water usage and weather patterns. By reducing water waste and enhancing irrigation efficiency, these technologies contribute to the sustainable use of water resources in agriculture. Similarly, systems that integrate soil sensors and AI, like Weenat's Europe-wide network, allow for precise monitoring of soil moisture levels, ensuring that irrigation is applied only when necessary. This technological innovation significantly reduces water consumption in agriculture, helping to address the challenges of water scarcity and climate change.

In addition to water management, the circular economy in the water-food nexus emphasizes the importance of recovering and reusing nutrients from agricultural systems. Phosphorus recovery through processes such as struvite crystallization, as demonstrated by Waternet's Airprex process in Amsterdam, provides a sustainable solution to reducing reliance on synthetic fertilizers while enhancing nutrient use efficiency. These nutrient recovery systems also offer significant economic benefits, reducing operational costs for wastewater treatment plants and creating valuable by-products like high-quality fertilizer. By promoting nutrient recycling, these practices contribute to the long-term sustainability of agriculture and support the circular economy's goal of closing resource loops.

Integrating renewable energy into agricultural systems further supports the circular economy within the water-food nexus. Projects such as Botswana's National Development Bank program for off-grid farmers, which promotes solar energy and water-efficient irrigation systems, demonstrate how renewable energy can enhance agricultural productivity while reducing reliance on fossil fuels. Similarly, biogas production from organic waste, as seen at the Rialto Bioenergy Facility in California, highlights the potential of circular waste management practices to generate renewable energy and produce valuable agricultural byproducts like fertilizer. These initiatives contribute to environmental sustainability and economic resilience by reducing greenhouse gas emissions and creating new revenue streams for farmers.

Finally, financing the circular economy in the water-food nexus requires innovative investment strategies and strong partnerships between the public and private sectors. Blended finance models, such as the Acumen Resilient Agriculture Fund, combine private capital with risk-tolerant funding from foundations and multilateral organizations to support climate-resilient agriculture. These models help de-risk investments in sustainable farming practices, enabling smallholder farmers to improve their resilience to climate change while attracting private investment. Public-private partnerships, such as the Nestlé and AWS collaboration, further demonstrate the value of cross-sectoral cooperation in achieving water sustainability goals. By leveraging financial resources and expertise from both sectors, these partnerships help scale up investments in water infrastructure, irrigation systems, and conservation practices.

In conclusion, the best practices outlined throughout this chapter demonstrate the significant potential of the circular economy to enhance water and food security while promoting environmental resilience. By reducing water usage, reusing and recycling water, recovering nutrients, and integrating renewable energy, regions worldwide can transition toward more sustainable water and food systems. These practices underscore the importance of adopting circular economy principles to address the growing global challenges of water scarcity, food insecurity, and environmental degradation. The successful implementation of these practices will depend on ongoing innovation, collaboration, and investment, with the potential to significantly improve agricultural productivity, ecosystem resilience, and water security.

Notes

[1]   "Feeding the World Sustainably," 2012, accessed September 13th, 2024, https://www.un.org/en/chronicle/article/feeding-world-sustainably.

[2]   Alberto Boretti and Lorenzo Rosa, "Reassessing the projections of the World Water Development Report," *npj Clean Water* 2, no. 1 (2019/07/31 2019), https://doi.org/10.1038/s41545-019-0039-9.

[3]   Xingcai Liu, et al., "Global agricultural water scarcity assessment incorporating blue and green water availability under future climate change," *Earth's Future* 10, no. 4 (2022/04/01 2022), https://doi.org/10.1029/2021EF002567.

[4]   Robert C. Brears, *The Green Economy and the Water-Energy-Food Nexus* (Springer International Publishing, 2023). https://books.google.co.nz/books?id=xDTYEAAAQBAJ.

[5]   "What is the Circular Economy?," 2024, accessed September 13th, 2024, https://iap.unido.org/articles/what-circular-economy.

[6]   "What is circular economy and why does it matter?," 2023, accessed September 13th, 2024, https://climatepromise.undp.org/news-and-stories/what-is-circular-economy-and-how-it-helps-fight-climate-change.

[7]   "Circular economy: definition, importance and benefits," 2023, accessed September 13th, 2024, https://www.europarl.europa.eu/topics/en/article/20151201STO05603/circular-economy-definition-importance-and-benefits.

[8]   "What is a circular economy?," 2024, accessed September 13th, 2024, https://www.ellenmacarthurfoundation.org/topics/circular-economy-introduction/overview.

[9]   "What is the circular economy?," 2023, accessed September 13th, 2024, https://www.chathamhouse.org/2021/06/what-circular-economy.

[10]   Robert C. Brears, *Natural Resource Management and the Circular Economy* (Cham, Switzerland: Springer International Publishing, 2018). https://books.google.co.nz/books?id=SphMDwAAQBAJ.

[11]   Robert C. Brears, "Circular Economy Cities," in *The Palgrave Encyclopedia of Urban and Regional Futures* (Cham: Springer International Publishing, 2020).

[12]   Robert C. Brears, "Circular Water Economy," in *The Palgrave Encyclopedia of Urban and Regional Futures*, ed. Robert C. Brears (Cham: Springer International Publishing, 2022).

[13]   Robert C. Brears, *Developing the Circular Water Economy* (Cham, Switzerland: Palgrave Macmillan, 2020).

[14]   Robert C. Brears, *Water Resources Management*, Innovative and Green Solutions, (De Gruyter, 2024). https://doi.org/10.1515/9783111028101.

[15]   Robert C. Brears, *Regional Water Security* (Wiley, 2021).

[16]   "Drought Resilience Innovation Grants," 2024, accessed September 13th, 2024, https://www.agriculture.gov.au/agriculture-land/farm-food-drought/drought/future-drought-fund/research-adoption-program/drought-resilience-innovation-grants.

[17]   "Water Irrigation Efficiency Program," 2024, accessed September 13th, 2024, https://extension.arizona.edu/water-irrigation-efficiency-program.

[18]   Water Research Commission, "Sustainable Use of Greywater in Small-Scale Agriculture and Gardens in South Africa," (2011). https://www.wrc.org.za/wp-content/uploads/mdocs/Rodda%20N.pdf.

[19]   Environment Agency – Abu Dhabi, "New water source," (2024). https://ead.gov.ae/-/media/Project/EAD/EAD/Documents/Resources/New-Water-Source-Factsheet-English.pdf.

[20]   University of Maryland Extension, "Putting Recycled Water to Work in Maryland Agriculture," 2024, no. September 13th (2024). https://extension.umd.edu/arec.umd.edu/sites/extension.umd.edu/files/publications/3.23.20%20Recycled%20Water%20to%20Work.pdf.

[21]   "Al Wathbah-2 wastewater treatment plant and Abu Dhabi irrigation scheme," 2024, accessed September 13th, 2024, https://ncwr.arabwatercouncil.org/reuse-of-treated-waste-water-success-stories/al-wathbah-2-wastewater-treatment-plant-and-abu-dhabi-irrigation-scheme/.

https://doi.org/10.1515/9783111341385-010

[22] "Good Soil. Better Earth," 2024, accessed September 13th, 2024, https://www.dcwater.com/about-dc-water/what-we-do/wastewater-treatment/biosolids.

[23] "Biosolids," 2024, accessed September 18th, 2024, https://www.unitedutilities.com/corporate/about-us/what-we-do/bioresources/biosolids/.

[24] "Conservation Grants," 2024, accessed September 13th, 2024, https://mda.maryland.gov/resource_conservation/Pages/financial_assistance.aspx.

[25] "Mulkear River Catchment Project," 2024, accessed September 13th, 2024, https://www.mulkear eip.com.

[26] William J. Cosgrove and Daniel P. Loucks, "Water management: Current and future challenges and research directions," *Water Resources Research* 51, no. 6 (2015/06/01 2015), https://doi.org/10.1002/2014WR016869.

[27] Pacific Institute, "The multiple benefits of water efficiency for California agriculture," (2014). https://pacinst.org/wp-content/uploads/2014/07/pacinst-water-efficiency-benefits-ca-ag.pdf.

[28] Gabrijel Ondrasek, "Water Scarcity and Water Stress in Agriculture," in *Physiological Mechanisms and Adaptation Strategies in Plants Under Changing Environment: Volume 1*, ed. Parvaiz Ahmad and Mohd Rafiq Wani (New York, NY: Springer, 2014).

[29] Carlo Ingrao, et al., "Water scarcity in agriculture: An overview of causes, impacts and approaches for reducing the risks," *Heliyon* 9, no. 8 (Aug 2023), https://doi.org/10.1016/j.heliyon.2023.e18507.

[30] Yu Chen, et al., "Water-saving techniques: physiological responses and regulatory mechanisms of crops," *Advanced Biotechnology* 1, no. 4 (2023/10/26 2023), https://doi.org/10.1007/s44307-023-00003-7.

[31] "Soil moisture conservation techniques," 2024, accessed September 9th, 2024, https://www.ctc-n.org/technologies/soil-moisture-conservation-techniques.

[32] Soumya R. Sahoo, et al., "Knowledge-based optimal irrigation scheduling of agro-hydrological systems," *Sustainability* 14, no. 3 (2022), https://doi.org/10.3390/su14031304.

[33] "Nutrient Management," 2024, accessed September 9th, 2024, https://www.farmers.gov/conservation/nutrient-management.

[34] "Conservation tillage," 2024, accessed September 9th, 2024, https://sarep.ucdavis.edu/sustainable-ag/conservation-tillage.

[35] "Water Harvesting and Storage," 2024, accessed September 9th, 2024, https://www.fao.org/land-water/water/water-management/water-storage/en/.

[36] Robert C. Brears, *Nature-based Solutions to 21st Century Challenges* (Oxfordshire, UK: Routledge, 2020). https://books.google.co.nz/books?id=0zgKzQEACAAJ.

[37] EIP-AGR, "Sustainable and resilient farming," (2020). https://ec.europa.eu/eip/agriculture/sites/default/files/eip-agri_brochure_agro-ecology_2020_en_web.pdf.

[38] "Considerations in Adopting Variable Rate Irrigation," 2017, accessed September 9th, 2024, https://water.unl.edu/article/agricultural-irrigation/considerations-adopting-variable-rate-irrigation.

[39] "Variable Rate Irrigation," 2024, accessed September 9th, 2024, https://www.irrigationnz.co.nz/PracticalResources/SpecialistSystems/PrecisionIrrigation.

[40] "Soil moisture monitoring," 2024, accessed September 9th, 2024, https://www.ctc-n.org/technologies/soil-moisture-monitoring.

[41] Hamlyn G. Jones, "Irrigation scheduling: advantages and pitfalls of plant-based methods," *Journal of Experimental Botany* 55, no. 407 (2004), https://doi.org/10.1093/jxb/erh213.

[42] Sandeep Bhatti, et al., "Toward automated irrigation management with integrated crop water stress index and spatial soil water balance," *Precision Agriculture* 24, no. 6 (2023/12/01 2023), https://doi.org/10.1007/s11119-023-10038-4.

[43] Weili Duan, et al., "Recent advancement in remote sensing technology for hydrology analysis and water resources management," *Remote Sensing* 13, no. 6 (2021), https://doi.org/10.3390/rs13061097.

[44] "Application of GIS In Water Resource Management," 2024, accessed September 9th, 2024, https://www.spatialpost.com/gis-in-water-resource-management/#google_vignette.

[45] Ravneet Kaur Sidhu, Ravinder Kumar and Prashant Singh Rana, "Machine learning based crop water demand forecasting using minimum climatological data," *Multimedia Tools and Applications* 79, no. 19 (2020/05/01 2020), https://doi.org/10.1007/s11042-019-08533-w.

[46] "Precision agriculture, AI, and water efficiency: The future of farming," Mark and Focus, 2023, accessed September 9th, 2024, https://medium.com/mark-and-focus/precision-agriculture-ai-and-water-efficiency-the-future-of-farming-b959ac0b6017.

[47] Jay Gohil, et al., "Advent of Big Data technology in environment and water management sector," *Environmental Science and Pollution Research* 28, no. 45 (2021/12/01 2021), https://doi.org/10.1007/s11356-021-14017-y.

[48] D. Maria Manuel Vianny, et al., "Water optimization technique for precision irrigation system using IoT and machine learning," *Sustainable Energy Technologies and Assessments* 52 (2022/08/01 2022), https://doi.org/https://doi.org/10.1016/j.seta.2022.102307, https://www.sciencedirect.com/science/article/pii/S2213138822003599.

[49] Samer El-Zahab and Tarek Zayed, "Leak detection in water distribution networks: an introductory overview," *Smart Water* 4, no. 1 (2019/06/11 2019), https://doi.org/10.1186/s40713-019-0017-x.

[50] Gerard Arbat and Daniele Masseroni, "The use and management of agricultural irrigation systems and technologies," *Agriculture* 14, no. 2 (2024), https://doi.org/10.3390/agriculture14020236.

[51] V. Sumanth et al., "Smart watering: Revolutionizing irrigation with AR and IoT" (paper presented at the Recent Advances in Civil Engineering for Sustainable Communities, Singapore, 2024// 2024).

[52] Kent Kovacs, et al., "On-farm reservoir adoption in the presence of spatially explicit groundwater use and recharge," *Journal of Agricultural and Resource Economics* 40, no. 1 (2015), http://www.jstor.org/stable/44131275.

[53] "Rainwater Harvesting – Artificial Recharge Of Groundwater in India," 2024, accessed September 9th, 2024, https://www.ceew.in/publications/sustainable-agriculture-india/rainwater-harvesting.

[54] Erion Bwambale and Felix K. Abagale, "Smart Irrigation Monitoring and Control," in *Encyclopedia of Smart Agriculture Technologies*, ed. Qin Zhang (Cham: Springer International Publishing, 2022).

[55] Simona Violino, et al., "A data-driven bibliometric review on precision irrigation," *Smart Agricultural Technology* 5 (2023/10/01 2023), https://doi.org/10.1016/j.atech.2023.100320, https://www.sciencedirect.com/science/article/pii/S2772375523001491.

[56] "Water management: Weenat is innovating to provide all agricultural branches with key data in the coming years," 2023, accessed September 9th, 2024, https://weenat.com/en/innovation-water-management-agriculture/.

[57] Bruno Da Silva, et al., "Satellite-based ET estimation using Landsat 8 images and SEBAL model," *Revista Ciencia Agronomica* 49 (04/01 2018).

[58] Timothy Foster, Taro Mieno and N. Brozović, "Satellite-based monitoring of irrigation water use: Assessing measurement errors and their implications for agricultural water management policy," *Water Resources Research* 56, no. 11 (2020/11/01 2020), https://doi.org/10.1029/2020WR028378.

[59] Prem Rajak, et al., "Internet of Things and smart sensors in agriculture: Scopes and challenges," *Journal of Agriculture and Food Research* 14 (2023/12/01 2023), https://doi.org/10.1016/j.jafr.2023.100776, https://www.sciencedirect.com/science/article/pii/S2666154323002831.

[60] Kamila Koteish, et al., "AGRO: A smart sensing and decision-making mechanism for real-time agriculture monitoring," *Journal of King Saud University – Computer and Information Sciences* 34, no. 9 (2022/10/01 2022), https://doi.org/10.1016/j.jksuci.2022.06.017, https://www.sciencedirect.com/science/article/pii/S1319157822002257.

[61] "Getting the measure of soil moisture in Australia," 2021, accessed September 9th, 2024, https://www.tern.org.au/news-smips-soil-moisture/.

[62] "Water markets," 2024, accessed September 19th, 2024, https://www.mdba.gov.au/water-use/water-markets.

[63] "Irrigation Efficiencies Grant Program (IEGP)," 2024, accessed September 19th, 2024, https://www.scc.wa.gov/programs/irrigation-efficiencies-grant-program-iegp.

[64] "Agricultural Water Use Efficiency," 2024, accessed September 9th, 2024, https://water.ca.gov/Programs/Water-Use-And-Efficiency/Agricultural-Water-Use-Efficiency.

[65] "Soil & Water – Resource Conservation Workshop," 2024, accessed September 19th, 2024, https://www.ncagr.gov/divisions/soil-water-conservation/programs-initiatives/education-programs/resource-conservation-workshop#ForDistricts-2549.

[66] Joseph, Philip M. Haygarth Holden, Jannette MacDonald, Alan Jenkins, Alison Sapiets, Harriet G. Orr, Nicola Dunn, Bob Harris, Phillippa L. Pearson, Dan McGonigle, Ann Humble, Martin Ross, Jim Harris, Theresa Meacham and Tim Benton. "Agriculture's impacts on water quality," (2015). https://nora.nerc.ac.uk/id/eprint/510550/.

[67] Huma Zia, et al., "The impact of agricultural activities on water quality: A case for collaborative catchment-scale management using integrated wireless sensor networks," *Computers and Electronics in Agriculture* 96 (2013/08/01 2013), https://doi.org/10.1016/j.compag.2013.05.001, https://www.sciencedirect.com/science/article/pii/S0168169913001063.

[68] "Chemical contamination and agriculture," 2024, accessed August 7th, 2024, https://niwa.co.nz/freshwater/kaitiaki-tools/what-impacts-interest-you/chemical-contamination/causes-chemical-contamination/chemical-contamination-and-agriculture#:~:text=Excess%20nutrients%20can%20also%20lead,which%20may%20lead%20to%20death.

[69] Food and Agriculture Organization of the United Nations and International Water Management Institute, (2023). https://openknowledge.fao.org/server/api/core/bitstreams/eb0ed8bb-545a-4d6b-84b2-d8c8bbd96b91/content.

[70] Nur Syabeera Begum Nasir Ahmad, et al., "A systematic review of soil erosion control practices on the agricultural land in Asia," *International Soil and Water Conservation Research* 8, no. 2 (2020/06/01 2020), https://doi.org/10.1016/j.iswcr.2020.04.001, https://www.sciencedirect.com/science/article/pii/S2095633920300216.

[71] Jaqueline Stenfert Kroese, et al., "Agricultural land is the main source of stream sediments after conversion of an African montane forest," *Scientific Reports* 10, no. 1 (2020/09/09 2020), https://doi.org/10.1038/s41598-020-71924-9.

[72] "The Causes and Effects of Soil Erosion, and How to Prevent It," 2020, accessed August 7th, 2024, https://www.wri.org/insights/causes-and-effects-soil-erosion-and-how-prevent-it.

[73] Michael F. Chislock, Enrique Doster, Rachel A. Zitomer and Alan E. Wilson, "Eutrophication: Causes, consequences, and controls in aquatic ecosystems," *Nature Education Knowledge* 4, no. 4 (2013).

[74] "What is Eutrophication? Understanding the Impact on Aquatic Ecosystems," 2024, accessed August 7th, 2024, https://www.americanoceans.org/facts/what-is-eutrophication/.

[75] US EPA, "Harmful algal blooms and drinking water," (2016). https://www.epa.gov/sites/default/files/2016-11/documents/harmful_algal_blooms_and_drinking_water_factsheet.pdf.

[76] "The Effects: Dead Zones and Harmful Algal Blooms," 2024, accessed August 7th, 2024, https://www.epa.gov/nutrientpollution/effects-dead-zones-and-harmful-algal-blooms.

[77] Kevin Parris, "Impact of agriculture on water pollution in OECD countries: Recent trends and future prospects," *International Journal of Water Resources Development* 27, no. 1 (2011/03/01 2011), https://doi.org/10.1080/07900627.2010.531898.

[78] "How to Cut Soil Erosion Risk in Indonesia," 2020, accessed August 7th, 2024, https://wri-indonesia.org/en/insights/how-cut-soil-erosion-risk-indonesia.

[79] "Sustainable Agriculture," 2024, accessed August 7th, 2024, https://www.nifa.usda.gov/topics/sustainable-agriculture.

[80] GIZ, "What is sustainable agriculture?," (2015). https://www.giz.de/en/downloads/giz2015-en-what-is-sustain-agric.pdf.

[81] "Sustainable Food and Agriculture," 2024, accessed August 7th, 2024, https://www.fao.org/sustain ability/en/.

[82] Sarah Velten, et al., "What is sustainable agriculture? A systematic review," *Sustainability* 7, no. 6 (2015), https://doi.org/10.3390/su7067833.

[83] Ahmad Bathaei and Dalia Štreimikienė, "Renewable energy and sustainable agriculture: Review of indicators," *Sustainability* 15, no. 19 (2023), https://doi.org/10.3390/su151914307.

[84] Xue Yang, et al., "Crop rotational diversity enhances soil microbiome network complexity and multifunctionality," *Geoderma* 436 (2023/08/01 2023), https://doi.org/10.1016/j.geoderma.2023. 116562, https://www.sciencedirect.com/science/article/pii/S0016706123002392.

[85] IUCN, "Crop diversification," (2023). https://www.iucn.org/sites/default/files/2023-05/practice_b. crop-diversification_final.pdf.

[86] Chang Liu, et al., "Chapter Six – Diversifying Crop Rotations Enhances Agroecosystem Services and Resilience," in *Advances in Agronomy*, ed. Donald L. Sparks (Academic Press, 2022).

[87] "Organic agriculture," 2024, accessed August 7th, 2024, https://www.fao.org/organicag/oa-faq/oa-faq1/en/.

[88] Ashoka Gamage, et al., "Role of organic farming for achieving sustainability in agriculture," *Farming System* 1, no. 1 (2023/04/01 2023), https://doi.org/10.1016/j.farsys.2023.100005, https://www.science direct.com/science/article/pii/S2949911923000059.

[89] Organics Europe, "Organic agriculture and its benefits for climate and biodiversity," (2022). https://www.organicseurope.bio/content/uploads/2022/04/IFOAMEU_advocacy_organic-benefits-for-climate-and-biodiversity_2022.pdf?dd.

[90] "Integrated Pest Management (IPM) Principles," 2024, https://www.epa.gov/safepestcontrol/inte grated-pest-management-ipm-principles#:~:text=Integrated%20Pest%20Management%20(IPM)% 20is,their%20interaction%20with%20the%20environment.

[91] "Integrated Pest Management (IPM)," 2024, accessed August 7th, 2024, https://sarep.ucdavis.edu/ sustainable-ag/ipm.

[92] "Pest and Pesticide Management," 2024, accessed August 7th, 2024, https://www.fao.org/pest-and-pesticide-management/ipm/integrated-pest-management/en/.

[93] Mutiu Abolanle Busari, et al., "Conservation tillage impacts on soil, crop and the environment," *International Soil and Water Conservation Research* 3, no. 2 (2015/06/01 2015), https://doi.org/10.1016/ j.iswcr.2015.05.002, http://www.sciencedirect.com/science/article/pii/S2095633915300630.

[94] "Conservation tillage," 2024, accessed August 7th, 2024, https://sarep.ucdavis.edu/sustainable-ag /conservation-tillage.

[95] Brooke Rust and John D. Williams, "How Tillage Affects Soil Erosion and Runoff," (2009). https://www.ars.usda.gov/ARSUserFiles/20740000/PublicResources/How%20Tillage%20Affects% 20Soil%20Erosion%20and%20Runoff.pdf.

[96] "Cover Crops for Sustainable Crop Rotations," 2024, accessed August 7th, 2024, https://www.sare. org/resources/cover-crops/.

[97] Environment Southland, "A guide to catch and cover crops," (2021). https://www.es.govt.nz/reposi tory/libraries/id:26gi9ayo517q9stt81sd/hierarchy/community/farming/good-management-practice /documents/Land%20sustainability%20guides%20and%20factsheets/A%20guide%20to%20catch% 20and%20cover%20crops.pdf.

[98] "Cover Crops and Crop Rotation," 2024, https://www.usda.gov/peoples-garden/soil-health/cover-crops-crop-rotation.

[99] Badege Bishaw, et al., "Agroforestry for sustainable production and resilient landscapes," *Agroforestry Systems* 96, no. 3 (2022/03/01 2022), https://doi.org/10.1007/s10457-022-00737-8.

[100] "Agroforestry," 2024, accessed August 7th, 2024, https://sarep.ucdavis.edu/are/ecosystem/agrofor estry.

[101] "Agroforestry," 2024, accessed August 7th, 2024, https://attra.ncat.org/topics/agroforestry/.

[102] Udayakumar Sekaran, et al., "Role of integrated crop-livestock systems in improving agriculture production and addressing food security – A review," *Journal of Agriculture and Food Research* 5 (2021/09/01 2021), https://doi.org/10.1016/j.jafr.2021.100190, https://www.sciencedirect.com/sci ence/article/pii/S2666154321000922.

[103] Bruno J. R. Alves, Beata E. Madari and Robert M. Boddey, "Integrated crop–livestock–forestry systems: prospects for a sustainable agricultural intensification," *Nutrient Cycling in Agroecosystems* 108, no. 1 (2017/05/01 2017), https://doi.org/10.1007/s10705-017-9851-0.

[104] Alan J. Franzluebbers, et al., "Toward agricultural sustainability through integrated crop–livestock systems. III. Social aspects," *Renewable Agriculture and Food Systems* 29, no. 3 (2014), https://doi.org/ 10.1017/S174217051400012X, https://www.cambridge.org/core/product/ B92271F14AC342DD78548479CE9F1F72.

[105] GOA, "Precision Agriculture: Benefits and Challenges for Technology Adoption and Use," (2024). https://www.gao.gov/products/gao-24-105962.

[106] "Benefits and Evolution of Precision Agriculture," 2024, accessed August 9th, 2024, https://www.ars. usda.gov/oc/utm/benefits-and-evolution-of-precision-agriculture/.

[107] "Climate-Smart Agriculture Program," 2024, accessed August 9th, 2024, https://www.agriculture. gov.au/agriculture-land/farm-food-drought/natural-resources/landcare/climate-smart.

[108] "MoFA establishes demonstration farms in districts," 2016, accessed August 9th, 2024, https://www. graphic.com.gh/news/general-news/mofa-establishes-demonstration-farms-in-districts.html.

[109] UNESCO World Water Assessment Programme, "Nature-based solutions (NBS) and water," (2018). https://unesdoc.unesco.org/ark:/48223/pf0000261605.

[110] FAO, "Nature-Based Solutions for agricultural water management and food security," (2018). https://openknowledge.fao.org/server/api/core/bitstreams/818a9e85-12e7-415a-b005 -05d37a33377d/content.

[111] Elisabeth Simelton, et al., "NBS framework for agricultural landscapes," Policy and practice reviews, *Frontiers in Environmental Science* 9 (2021-August-05 2021), https://doi.org/10.3389/fenvs.2021. 678367, https://www.frontiersin.org/journals/environmental-science/articles/10.3389/fenvs.2021. 678367.

[112] "Constructed Farm Wetlands," 2024, accessed August 9th, 2024, https://www.wwt.org.uk/our-work /wetland-conservation-unit/resources/constructed-farm-wetlands.

[113] "AGROforestry and MIXed farming systems – Participatory research to drive the transition to a resilient and efficient land use in Europe," 2024, accessed August 9th, 2024, https://eu-cap-network. ec.europa.eu/projects/agroforestry-and-mixed-farming-systems-participatory-research-drive- transition-resilient_de.

[114] "Water Recycling and Reuse in Agriculture for Circularity of Food and Water," Water-Food Nexus, 2023, accessed September 5th, 2024, https://medium.com/water-food-nexus/water-recycling-and- reuse-in-agriculture-for-circularity-of-food-and-water-f08fe4b131b3#:~:text=In%20conclusion%2C% 20water%20recycling%20and,increasing%20demand%20and%20changing%20climate.

[115] "A circular economy for food will help people and nature thrive," 2024, accessed September 5th, 2024, https://www.ellenmacarthurfoundation.org/topics/food/overview.

[116] "Food and the circular economy – deep dive," 2024, accessed September 5th, 2024, https://www. ellenmacarthurfoundation.org/food-and-the-circular-economy-deep-dive.

[117] H. Chen, J. M. Burke and E. E. Prepas, "Cyanobacterial Toxins in Fresh Waters," in *Encyclopedia of Environmental Health*, ed. Jerome O. Nriagu (Burlington: Elsevier, 2011).

[118] Fabio Masi, et al., "Chapter Four – Possibilities of Nature-Based and Hybrid Decentralized Solutions for Reclaimed Water Reuse," in *Advances in Chemical Pollution, Environmental Management and Protection*, ed. Paola Verlicchi (Elsevier, 2020).

[119] E. Obotey Ezugbe and Sudesh Rathilal, "Membrane technologies in wastewater treatment: A review," *Membranes (Basel)* 10, no. 5 (Apr 30 2020), https://doi.org/10.3390/membranes10050089.

[120] Robert C. Brears, *Urban Water Security* (Chichester, UK; Hoboken, NJ: John Wiley & Sons, 2016).

[121] "Recycled water networks," 2024, accessed September 5th, 2024, https://www.sawater.com.au/water-and-the-environment/how-we-deliver-your-water-services/recycling-and-reuse-network.

[122] "Northern Adelaide Irrigation Scheme," 2024, accessed September 4th, 2024, https://www.sawater.com.au/nais.

[123] UNEP Copenhagen Climate Centre, "Drip irrigation systems fed by rainwater harvesting," (2022). https://tech-action.unepccc.org/wp-content/uploads/sites/2/2022/02/agriculture-sector-policy-brief-jamaica.pdf.

[124] Waternet, "10 years of Phosphorus Recovery at WWTP Amsterdam West," (2023). https://www.stowa.nl/sites/default/files/2023-06/2%20Alex%20Veltman%20%28Waternet%29%20Airprex%20Amsterdam%20West.pdf.

[125] Liliana Cifuentes-Torres, Gabriel Correa-Reyes and Leopoldo G. Mendoza-Espinosa, "Can reclaimed water be used for sustainable food production in aquaponics?," Hypothesis and Theory, *Frontiers in Plant Science* 12 (2021-June-04 2021), https://doi.org/10.3389/fpls.2021.669984, https://www.frontiersin.org/journals/plant-science/articles/10.3389/fpls.2021.669984.

[126] USDA, "Less is more: eco-intensification using recycled drainage water for fertigation," (2022). https://portal.nifa.usda.gov/web/crisprojectpages/1027913-less-is-more-eco-intensification-using-recycled-drainage-water-for-fertigation.html.

[127] "NASA Research Launches a New Generation of Indoor Farming," 2021, accessed September 5th, 2024, https://spinoff.nasa.gov/indoor-farming.

[128] "Organic Matter and Soil Amendments," 2023, accessed September 5th, 2024, https://extension.umd.edu/resource/organic-matter-and-soil-amendments/.

[129] María F. Jaramillo and Inés Restrepo, "Wastewater reuse in agriculture: A review about its limitations and benefits," *Sustainability* 9, no. 10 (2017), https://doi.org/10.3390/su9101734.

[130] Simon Toze, "Reuse of effluent water – benefits and risks," *Agricultural Water Management* 80, no. 1 (2006/02/24 2006), https://doi.org/10.1016/j.agwat.2005.07.010, https://www.sciencedirect.com/science/article/pii/S0378377405002957.

[131] John Anderson, "The environmental benefits of water recycling and reuse," *Water Supply* 3, no. 4 (2003), https://doi.org/10.2166/ws.2003.0041.

[132] "Hydroponics," 2024, accessed September 4th, 2024, https://www.nal.usda.gov/farms-and-agricultural-production-systems/hydroponics.

[133] Jung Eek Son, Hak Jin Kim and Tae In Ahn, "Chapter 20 – Hydroponic Systems," in *Plant Factory (Second Edition)*, ed. Toyoki Kozai, Genhua Niu and Michiko Takagaki (Academic Press, 2020).

[134] "Types Of Hydroponic Systems," 2024, accessed September 4th, 2024, https://getgrowee.com/types-of-hydroponic-systems/.

[135] "Water Use Efficiency in Hydroponics and Aquaponics," 2024, accessed September 4th, 2024, https://zipgrow.com/water-use-efficiency-hydroponics-aquaponics/.

[136] "Hydroponics vs. Aquaponics – A Complete, and Honest Comparison," 2024, accessed September 4th, 2024, https://www.trees.com/gardening-and-landscaping/hydroponics-vs-aquaponics.

[137] "Urban Farming and Water: Innovations for Sustainable Agriculture," 2023, accessed August 30th, 2024, https://medium.com/mark-and-focus/urban-farming-and-water-innovations-for-sustainable-agriculture-da4732e00730.

[138] "Vertical Farming – No Longer A Futuristic Concept," 2024, accessed September 4th, 2024, https://www.ars.usda.gov/oc/utm/vertical-farming-no-longer-a-futuristic-concept/.

[139] S. H. van Delden, et al., "Current status and future challenges in implementing and upscaling vertical farming systems," *Nature Food* 2, no. 12 (2021/12/01 2021), https://doi.org/10.1038/s43016-021-00402-w.

[140] "Vertical Farming: A Sustainable Solution for Climate Resilience and Resource Optimization," Mark and Focus, 2024, accessed September 4th, 2024, https://medium.com/global-climate-solutions/vertical-farming-a-sustainable-solution-for-climate-resilience-and-resource-optimization-f568182415a4.

[141] Dafni Despoina Avgoustaki and George Xydis, "How energy innovation in indoor vertical farming can improve food security, sustainability, and food safety?," no. 2452-2635 (Print).

[142] ""GigaFarm" capable of replacing 1% of UAE food imports set for construction in Dubai Food Tech Valley," 2023, accessed September 19th, 2024, https://www.intelligentgrowthsolutions.com/press-release/gigafarm-announcement-in-dubai.

[143] Jonas Jägermeyr, et al., "Climate impacts on global agriculture emerge earlier in new generation of climate and crop models," *Nature Food* 2, no. 11 (2021/11/01 2021), https://doi.org/10.1038/s43016-021-00400-y.

[144] Amman KhokharVoytas, et al., "Genetic modification strategies for enhancing plant resilience to abiotic stresses in the context of climate change," *Functional & Integrative Genomics* 23, no. 3 (2023/08/29 2023), https://doi.org/10.1007/s10142-023-01202-0.

[145] Syed Shan-e-Ali Zaidi, et al., "Engineering crops of the future: CRISPR approaches to develop climate-resilient and disease-resistant plants," *Genome Biology* 21, no. 1 (2020/11/30 2020), https://doi.org/10.1186/s13059-020-02204-y.

[146] Amitava Aich, Dipayan Dey and Arindam Roy, "Climate change resilient agricultural practices: A learning experience from indigenous communities over India," *PLOS Sustainability and Transformation* 1, no. 7 (2022), https://doi.org/10.1371/journal.pstr.0000022.

[147] Márta Ladányi, Anna Divéky-Ertsey and László Csambalik, "Editorial: Adaptation of traditional crop cultivars to climate change in terms of nutritional aspects," Editorial, *Frontiers in Nutrition* 11 (2024-May-23 2024), https://doi.org/10.3389/fnut.2024.1427068, https://www.frontiersin.org/journals/nutrition/articles/10.3389/fnut.2024.1427068.

[148] Jiao Liu, et al., "Association between reactive oxygen species, transcription factors, and candidate genes in drought-resistant sorghum," *International Journal of Molecular Sciences* 25, no. 12 (2024), https://doi.org/10.3390/ijms25126464.

[149] "Drought-Resilient Crops: Farmers' New Allies," 2024, accessed September 4th, 2024, https://thefarminginsider.com/drought-resilient-crops-farmers/.

[150] Abel Ruiz-Giralt, et al., "Small-scale farming in drylands: New models for resilient practices of millet and sorghum cultivation," *PLOS ONE* 18, no. 2 (2023), https://doi.org/10.1371/journal.pone.0268120.

[151] "Drought-Resilient Crops: A Guide," 2024, accessed September 4th, 2024, https://thefarminginsider.com/drought-resilient-crops/.

[152] "What is genetic modification (GM) of crops and how is it done?," 2016, accessed September 4th, 2024, https://royalsociety.org/news-resources/projects/gm-plants/what-is-gm-and-how-is-it-done/.

[153] Theresa Phillips, "Genetically modified organisms (GMOs): Transgenic crops and recombinant DNA technology," *Nature Education* 1, no. 1 (2008).

[154] "How GMO Crops Impact Our World," 2024, accessed September 4th, 2024, https://www.fda.gov/food/agricultural-biotechnology/how-gmo-crops-impact-our-world.

[155] Logayn T. Abushal, et al., "Agricultural biotechnology: Revealing insights about ethical concerns," *Journal of Biosciences* 46, no. 3 (2021/08/12 2021), https://doi.org/10.1007/s12038-021-00203-0.

[156] "Selective Breeding and Genetic Engineering," 2024, accessed September 4th, 2024, https://bio.libretexts.org/Courses/University_of_Pittsburgh/Environmental_Science_%28Whittinghill%29/13%3A_Agriculture/13.05%3A_Selective_Breeding_and_Genetic_Engineering.

[157] Muhammad Afzal, et al., "Potential breeding strategies for improving salt tolerance in crop plants," *Journal of Plant Growth Regulation* 42, no. 6 (2023/06/01 2023), https://doi.org/10.1007/s00344-022-10797-w.

[158] "Ancient environmental DNA provides solutions for global food security challenges," 2024, accessed September 4th, 2024, https://www.ebi.ac.uk/about/news/announcements/ancient-environmental-dna-provides-solutions-for-global-food-security-challenges/.

[159] Ritika B. Yadav, "Biodegradable Packaging: Recent Advances and Applications in Food Industry," in *Food Process Engineering and Technology: Safety, Packaging, Nanotechnologies and Human Health*, ed. Junaid Ahmad Malik, Megh R. Goyal and Anu Kumari (Singapore: Springer Nature Singapore, 2023).

[160] Helen Onyeaka, et al., "Current research and applications of starch-based biodegradable films for food packaging," *Polymers* 14, no. 6 (2022), https://doi.org/10.3390/polym14061126.

[161] "Food and beverage packaging: The rise of breathable packaging," 2023, accessed September 4th, 2024, https://www.foodinfotech.com/food-and-beverage-packaging-the-rise-of-breathable-packaging/.

[162] Mary R. Yan, Sally Hsieh and Norberto Ricacho, "Innovative food packaging, food quality and safety, and consumer perspectives," *Processes* 10, no. 4 (2022), https://doi.org/10.3390/pr10040747.

[163] "About us," 2024, accessed September 4th, 2024, https://lovefoodhatewaste.co.nz/about-us/.

[164] "Green energy transition for sustainable agriculture," 2024, accessed September 19th, 2024, https://iki-small-grants.de/k2project/green-energy-transition-for-sustainable-agriculture/.

[165] "Converting Food Waste to Bioenergy in Rialto," 2020, accessed September 4th, 2024, https://www.caclimateinvestments.ca.gov/2020-profiles/organics-grants.

[166] Frank Lohrberg, Lilli Lička, Lionella Scazzosi and Axel Timpe, "Urban Agriculture Europe," (2016). https://www.ideabooks.it/wp-content/uploads/2016/12/Urban-Agriculture-Europe.pdf.

[167] "What Is Urban Farming? Understanding Urban Agriculture," 2024, accessed August 30, 2024, https://unity.edu/careers/what-is-urban-farming/.

[168] "Urban Agriculture," 2024, accessed August 30, 2024, https://www.nal.usda.gov/farms-and-agricultural-production-systems/urban-agriculture.

[169] EFUA, "A typology of Urban Agriculture," (2022). https://www.efua.eu/sites/default/files/2022-10/3750426803_A%20typology%20of%20Urban%20Agriculture%2004102022%20Small_compressed.pdf.

[170] RUAF, "Urban Agriculture types/production systems and short food chains," (2023). https://ruaf.org/assets/2019/11/Module-3-Urban-Agriculture-types-production-systems-and-short-food-chains.pdf.

[171] "Urban Farming of the Future," Mark and Focus, 2022, accessed August 30, 2024, https://medium.com/mark-and-focus/urban-farming-of-the-future-ba6a53d3878e.

[172] "Urban Farming Ultimate Guide and Examples," 2024, accessed August 30th, 2024, https://grocycle.com/urban-farming/.

[173] "Amid crisis in food waste, celebrity chef Massimo Bottura gives produce a second life," 2023, accessed August 30th, 2024, https://www.unep.org/news-and-stories/video/amid-crisis-food-waste-celebrity-chef-massimo-bottura-gives-produce-second#:~:text=Every%20year%2C%20570%20million%20tons,million%20people%20globally%20go%20hungry.

[174] Juan Garzón, et al., "Systematic review of technology in aeroponics: Introducing the technology adoption and integration in sustainable agriculture model," *Agronomy* 13, no. 10 (2023), https://doi.org/10.3390/agronomy13102517.

[175] Rui D. Sousa, et al., "Challenges and solutions for sustainable food systems: The potential of home hydroponics," *Sustainability* 16, no. 2 (2024), https://doi.org/10.3390/su16020817.

[176] Robert C. Brears, *Blue and Green Cities: The Role of Blue-Green Infrastructure in Managing Urban Water Resources* (Springer International Publishing, 2023). https://books.google.co.nz/books?id=du3aEAAAQBAJ.

[177] "Rainwater Harvesting for Urban Agriculture," 2024, accessed September 2nd, 2024, https://refloh2o.com/rainwater-harvesting.

[178] "Composting: Solution to Food Loss and Waste," 2023, accessed September 2nd, 2024, https://www.unep.org/ietc/news/story/composting-solution-food-loss-and-waste.

[179] "Composting," 2024, accessed September 2nd, 2024, https://www.epa.gov/sustainable-management-food/composting.

[180] "Moisture In Compost: Everything You Need To Know," 2024, accessed September 2nd, 2024, https://www.compostmagazine.com/moisture-in-compost/#google_vignette.

[181] "Anaerobic Digestion (Small-scale)," 2024, accessed September 2nd, 2024, https://sswm.info/arctic-wash/module-4-technology/further-resources-wastewater-treatment/anaerobic-digestion-%28small-scale%29.

[182] "Methane digesters," 2024, accessed September 2nd, 2024, https://drawdown.org/solutions/methane-digesters.

[183] "Optimizing Biogas Feedstock for Maximum Efficiency," 2024, accessed September 2nd, 2024, https://www.nvend.io/blog/optimizing-biogas-feedstock-for-maximum-efficiency.

[184] Sidahmed Sidi Habib, et al., "Optimization of the factors affecting biogas production using the Taguchi design of experiment method," *Biomass* 4, no. 3 (2024), https://doi.org/10.3390/biomass4030038.

[185] "Parameters and Process Optimisation for Biogas," 2024, accessed September 2nd, 2024, https://energypedia.info/wiki/Parameters_and_Process_Optimisation_for_Biogas.

[186] "Food waste," 2024, accessed September 2nd, 2024, https://hallbart.stockholm/en/Sustainable-recycling/matavfall/.

[187] "Urban Farms And Food Forests As Green Infrastructure?," 2023, accessed September 2nd, 2024, https://www.biocycle.net/urban-farms-food-forests/.

[188] Megan Horst, Nathan McClintock and Lesli Hoey, "The intersection of planning, urban agriculture, and food justice: A review of the literature," *Journal of the American Planning Association* 83, no. 3 (2017/07/03 2017), https://doi.org/10.1080/01944363.2017.1322914.

[189] PolicyLink, "Growing Urban Agriculture: Equitable Strategies and Policies for Improving Access to Healthy Food and Revitalizing Communities," (2012). https://www.policylink.org/sites/default/files/URBAN_AG_FULLREPORT.PDF.

[190] "Urban Agriculture," 2024, accessed September 2nd, 2024, https://www.planning.org/knowledgebase/urbanagriculture/.

[191] Lydia Oberholtzer, Carolyn Dimitri and Andrew Pressman, "Urban agriculture in the United States: Characteristics, challenges, and technical assistance needs," *The Journal of Extension* 52, no. 6 (2014).

[192] UNEP and International Resource Panel, "Urban Agriculture's Potential to Advance Multiple Sustainability Goals," (2022). https://wedocs.unep.org/bitstream/handle/20.500.11822/38399/urban_agriculture_pol.pdf.

[193] "Growing urban agriculture," Stanford Social Innovation Review, 2017, https://ssir.org/articles/entry/growing_urban_agriculture.

[194] "Urban Agriculture Infrastructure and Operations Grant," 2024, accessed September 2nd, 2024, https://doee.dc.gov/service/urbag-infrastructure-operations-grant.

[195] Israel R. Orimoloye, "Water, energy and food nexus: Policy relevance and challenges," Systematic review, *Frontiers in Sustainable Food Systems* 5 (2022-February-01 2022), https://doi.org/10.3389/fsufs.2021.824322, https://www.frontiersin.org/journals/sustainable-food-systems/articles/10.3389/fsufs.2021.824322.

[196] "Catalyzing Private Sector Investment Through a Water Nexus Approach," 2023, accessed August 22nd, 2024, https://ndcpartnership.org/news/catalyzing-private-sector-investment-through-water-nexus-approach.

[197] "Agricultural Development," 2024, accessed August 22nd, 2024, https://www.gatesfoundation.org/our-work/programs/global-growth-and-opportunity/agricultural-development.

[198] "Agribusiness & Forestry," 2024, accessed August 22nd, 2024, https://www.ifc.org/en/what-we-do/sector-expertise/agribusiness-forestry.

[199] EIB, "The EIB Water Sector Fund," (2022). https://www.eib.org/en/publications/20220144-eib-water-sector-fund.

[200] Mohamed Behnassi, et al., "The Water, Climate, and Food Nexus: Linkages, Challenges and Emerging Solutions – An Introduction," in *The Water, Climate, and Food Nexus: Linkages, Challenges and Emerging Solutions*, ed. Mohamed Behnassi et al. (Cham: Springer International Publishing, 2024).

[201] APN, "Governing the Water-Energy-Food Nexus Approach for Creating Synergies and Managing Trade-offs," (2018). https://www.apn-gcr.org/publication/governing-the-water-energy-food-nexus-approach-for-creating-synergies-and-managing-trade-offs/.

[202] "Blended finance: How setting up a financial intermediary can accelerate sustainable development," 2023, accessed August 22nd, 2024, https://www.weforum.org/agenda/2023/04/blended-finance-financial-intermediation-can-accelerate-sustainable-development/.

[203] Pritee Sharma and Salla Nithyanth Kumar, "The global governance of water, energy, and food nexus: allocation and access for competing demands," *International Environmental Agreements: Politics, Law and Economics* 20, no. 2 (2020/06/01 2020), https://doi.org/10.1007/s10784-020-09488-2.

[204] Andrianto Ansari, et al., "Optimizing water-energy-food nexus: achieving economic prosperity and environmental sustainability in agriculture," Perspective, *Frontiers in Sustainable Food Systems* 7 (2023-August-29 2023), https://doi.org/10.3389/fsufs.2023.1207197, https://www.frontiersin.org/journals/sustainable-food-systems/articles/10.3389/fsufs.2023.1207197.

[205] Fabio Natalucci, Bo Li and Prasad Ananthakrishnan, "How blended finance can support climate transition in emerging and developing economies," *IMF Blog*, 2022, https://www.imf.org/en/Blogs/Articles/2022/11/15/how-blended-finance-can-support-climate-transition-in-emerging-and-developing-economies.

[206] "Agriculture Water Use Efficiency CDFA-DWR," 2024, accessed August 22nd, 2024, https://water.ca.gov/Work-With-Us/Grants-And-Loans/Agriculture-Water-Use-Efficiency-CDFA-DWR.

[207] "Climate-Resilient Irrigation," 2024, accessed August 22nd, 2024, https://www.worldbank.org/en/topic/climate-resilient-irrigation.

[208] Sebastián Sosa, Kristalina Georgieva and Björn Rother, "Global food crisis demands support for people, open trade, bigger local harvests," *IMF Blog*, 2022, https://www.imf.org/en/Blogs/Articles/2022/09/30/global-food-crisis-demands-support-for-people-open-trade-bigger-local-harvests.

[209] GCF, "Sectoral guide: Water security," (2022). https://www.greenclimate.fund/document/sectoral-guide-water-security.

[210] "Feed the Future," 2024, accessed August 22nd, 2024, https://www.usaid.gov/feed-the-future#:~:text=Feed%20the%20Future%20is%20America's%20initiative%20to%20combat,countries%20with%20great%20need%20and%20opportunity%20for%20improvement.

[211] "Activities in Kenya," 2024, accessed August 22nd, 2024, https://www.jica.go.jp/Resource/kenya/english/activities/activitiy01.html.

[212] "IFAD and JICA Renew their Memorandum of Cooperation for Rural Development, Food and Nutrition Security," 2024, accessed August 22nd, 2024, https://www.ifad.org/en/web/latest/-/ifad-and-jica-renew-their-memorandum-of-cooperation-for-rural-development-food-and-nutrition-security.

[213] "Climate Change Adaptation in Rural Areas of India (CCA-RAI)," 2024, accessed August 22nd, 2024, https://www.giz.de/en/worldwide/16603.html.

[214] "Venture Capital in Water: Key Transactions and Trends," 2017, accessed August 30th, 2024, https://www.bluefieldresearch.com/research/venture-capital-water/.

[215] Yongjie Zhang, Qiaoran Meng and Dayong Liu, "Venture capital and technology commercialization: evidence from China," *The Journal of Technology Transfer* (2024/02/10 2024), https://doi.org/10.1007/s10961-024-10063-z.

[216] Robert C. Brears, *Financing Nature-Based Solutions: Exploring Public, Private, and Blended Finance Models and Case Studies* (Springer International Publishing, 2022). https://books.google.co.nz/books?id=isNbEAAAQBAJ.

[217] Robert C. Brears, *Financing Water Security and Green Growth* (Oxford University Press, 2023), https://doi.org/10.1093/oso/9780192847843.001.0001.

[218] "Green Bond Principles (GBP)," 2024, accessed August 30th, 2024, https://www.icmagroup.org/sustainable-finance/the-principles-guidelines-and-handbooks/green-bond-principles-gbp/.

[219] "The Standard," 2024, accessed August 30th, 2024, https://www.climatebonds.net/standard/the-standard.

[220] "Acumen Resilient Agriculture Fund: Investing in Smallholder Farmers," 2021, accessed August 30th, 2024, https://acumen.org/news/acumen-resilient-agriculture-fund-investing-in-smallholder-farmers/.

[221] "Results-based financing: A potential game-changer for IFAD's future operations," 2024, accessed August 30th, 2024, https://www.ifad.org/en/web/latest/-/blog/results-based-financing-a-potential-game-changer-for-ifad-s-future-operations.

[222] "The AgriFI Kenya Challenge Fund," 2024, accessed August 30th, 2024, https://agrifichallengefund.org/about-the-agrifi-kenya-challenge-fund/.

[223] "Nestlé & Alliance for Water Stewardship Join Efforts for Water Usage," 2020, accessed August 30th, 2024, https://www.nestle-mena.com/en/media/pr/nestle-global-alliance-water-stewardship.

Index

https://doi.org/10.1515/9783111341385-011

www.ingramcontent.com/pod-product-compliance
Lightning Source LLC
Chambersburg PA
CBHW081530220326
41598CB00036B/6389